SAGE

ICRP Supporting Guidance 4

Annals of the ICRP

Development of the Draft 2005 recommendations of the ICRP: a collection of papers

ICRP Supporting Guidance 4

CONTENTS

SAGE

ICRP Supporting Guidance 4

Annals of the ICRP

Editorial

WHY DO THINGS HAVE TO CHANGE?

After all, a reasonable degree of stability is a very human desire, and usually very rational as well. However, the world is not totally static, neither are we – and it certainly would not do to lull ourselves into believing that when 14 years ago the International Commission on Radiological Protection adopted its current fundamental Recommendations (*Publication 60*), we had achieved perfection. There is always scope for improvement.

In fact, it has always been the Commission's policy to issue its Recommendations 'until further notice', usually replacing these at intervals of about 15 years. The current, '1990', Recommendations were developed during the 1980's, and thus in the autumn of 1997 the Commission decided to launch a project to 'consolidate or re-capitulate *Publication 60*'.

That decision was greeted with some surprise in some quarters. Had not *Publication 60* been integrated in national legislation quite recently in many countries? Yes, exactly – that is the point. Because of the long time frames required to produce new Recommendations, then issue the necessary supplementary guidance, and then go through all the steps to achieve the corresponding updates to legislation, it was necessary to set the wheels in motion in 1997. This was done in the hope of adopting the recommendations around 2005 and looking for legislation a few years later – i.e., some 15 years after the previous update.

An initial Commission discussion in the spring of 1998 pinpointed some of the issues to be addressed, first of all of course that which is implicit in the project heading: whether to start an unprejudiced discussion or simply to repeat the existing recommendations in proof of their continued validity – the former route was chosen. Some of the many other issues mentioned were:

• to state more clearly the ethical principles on which the Commission bases its work (not primarily to recommend specific ethics, but to clarify the inevitable value judgements inherent in risk management, thus also permitting others to apply alternative values if desired);
• to review the array of dosimetric quantities used (taking account of emerging evidence that both radiation weighting factors and tissue weighting factors needed some updating, and addressing some quantity denominations that have proved difficult to translate distinctly into other languages);
• to discuss critically the concept of a dose limit for members of the public (which is a very theoretical tool that can really only be used by national reg-

ulators and as guidelines for regulatory modelling, while practical legal limit-ation must be aimed at sources and dose rate and activity at the point of release); and

• to discuss thoroughly the biological basis of the recommendations (the linear no-threshold hypothesis regarded by the Commission as the most plausible current approximation of the actual dose-response at low dose and low dose-rate has been questioned by various lobbies for being either over-cautious or too lax, while phe-nomena such as genomic instability, age-dependence of risk, non-cancer endpoints, and genetic susceptibility variations all need to be discussed).

While a review of the existing recommendations was thus a timely and expected exercise, the procedure has evolved significantly to become much more interactive than in earlier times. The transition began when *Publication 60*, the 1990 Re-commendations, were prepared: a draft of those Recommendations was circulated widely for consultation before final adoption. This time, the approach is even more radical, involving the radiological protection community and indeed the general public from first principles.

Thus, the upshot of the spring 1998 discussion within the Commission was that Professor Roger Clarke was asked to write down his and other members' ideas as voiced at that meeting, not only as a contribution to the Commission's internal debate, but also as a paper for publication in the open literature. That initial paper, which initiated the public discussion about the next ICRP Recommenda-tions, is the first of three papers reproduced here by kind permission of the *Journal of Radiological Protection* (*JRP*) and the Institute of Physics Publishing. All of these papers have also been made available at no cost to all interested parties through the Internet, a mechanism without which it simply would not have been technically or economically feasible for ICRP to go for full-scale consultation.

Through the kind assistance of IRPA, the International Radiation Protection Association, and the various national organisations comprising IRPA, these initial proposals from Professor Clarke to ICRP were thoroughly analysed. At the IRPA X Congress in 2000, they were discussed, and a series of most helpful and constructive comments forwarded to the Commission have had quite a significant influence on the further development of this project. Several such comments were published as papers in their own right. Many international and national organisations have devoted considerable energy to participation in the debate; in particular, the Nuclear Energy Agency of the OECD has organised and co-organised with ICRP several workshops to refine and improve ideas.

The Commission's decision to use an open process has been welcomed and approved across the board. Some of the initial ideas were taken on board directly while others generated questions and more critical comments, which is as it should be and precisely the desired and expected result of consultation. A small number of comments were expressed in rather a loud voice and a few of them criticised views never proposed by the Commission or political justification decisions taken by the appropriate authorities in individual cases; even these less encouraging comments were quite useful in that they helped us to clarify the Commission's role and proposals.

In response to the initial proposal from Professor Clarke and the very comprehensive set of comments provided by the radiological protection community, the Commission itself issued the progress report reproduced here as the second *JRP* paper. This, in turn, was discussed in detail at scientific congresses and workshops in the four corners of the world, generating further comments and suggested amendments to the Commission's proposals. Again, this generated further development within ICRP. The third *JRP* paper reproduced here describes those developments and explains more clearly some of the reasons for making this revision now.

After this extensive process of iterative consultation on the conceptual ideas, the Commission was able to produce a draft actual text for the next set of fundamental ICRP Recommendations. This was launched at the IRPA XI Congress in 2004 and the summary of the draft is reproduced here; the full text was made available for downloading, at no cost of course, from our web site, www.icrp.org . At the time of writing, a full international and public consultation on that draft has just started; it will continue until the end of the year 2004.

Of course, the consultation will generate some critical comments and some further amendments of the text before final adoption – that is the very purpose of consultation. The comments and helpful criticism already received at the conceptual stage ensures that the present proposal is soundly based and reasonably well developed; the enormous interest and the very constructive nature of those comments convinces us that the comments now to be received on the actual text will be highly important, very useful, and always encouraging even when not concurring.

After the end of the consultation period, the Commission will of course need to spend considerable effort to ensure that all comments are evaluated and taken into account as due. It is hoped that we will be able to adopt a final amended version during 2005, with publication perhaps early in 2006 (this would be similar to *Publication 60*, known to everybody as the 1990 Recommendations because they were adopted in 1990, although they were printed only in 1991).

A consultation process of this kind requires much effort and is sometimes quite taxing for those providing comments. The very process can cause confusion because possible concepts are aired but then sometimes retracted when comments provide convincing arguments against those ideas. The Commission owes a great deal to all those experts and laymen in countries all over who have selflessly spent so much of their time and effort to help us 'get it right'. We hope and believe that the end result, in the shape of well discussed and developed Recommendations on Radiological Protection, will show that their effort was well worth their while.

JACK VALENTIN

PREFACE

In 1997, the International Commission on Radiological Protection (ICRP) decided to launch a project 'to consolidate or re-iterate *Publication 60* (ICRP, 1991)'. From then on, the project has been a standing item on the agenda of all Commission and Committee meetings, and will continue to be so until a final version of the next Recommendations is adopted, tentatively in the autumn of 2005.

Initially, while at the planning stage, the project was conducted by the Commission without any specific Task Group. A core Task Group was established formally in 2000, consisting of the 2001–2005 officers of the Commission (R.H. Clarke, L.-E. Holm, R. Cox, C. Streffer, F.A. Mettler, and B. Winkler, replaced in 2003 by A. Sugier), assisted by the Scientific Secretary (J. Valentin). This core group met in 2001 and informally at later occasions. However, for practical purposes, the entire Main Commission has acted as an extended Task Group working on this project. All of the four standing Committees of ICRP have participated in the manner of reference groups for the project.

The Terms of Reference of the project were to draft a next set of fundamental Recommendation of the ICRP, intended to replace the 1990 Recommendations in *Publication 60*.

Because of the importance and potential consequences of this major undertaking, it was decided at an early stage to consult widely and iteratively, first on conceptual ideas and then on more detailed proposals.

As a first step, the Commission requested R.H. Clarke to write a paper summarising proposals he and other Commission members had provided as early input, and to publish these proposal as a report in the open literature. That report (Clarke, 1999) was published in the *Journal of Radiological Protection* and is reproduced in the present issue of the *Annals of the ICRP*.

A second paper in the *Journal of Radiological Protection* (ICRP, 2001), also reproduced in this issue of the *Annals of the ICRP*, summarised the Commission's standpoint after discussion of the proposals in Clarke (1999) and the extensive comments received through consultation on those proposals.

Further discussions within the Commission and in the radiological protection community at large precipitated a third paper in the *Journal of Radiological Protection* (ICRP, 2003), again reproduced here. That paper provided a further progress report and described in some detail the reasons for issuing a next set of Recommendations replacing *Publication 60*.

Based on these developments and the extensive consultations so far, a proposed text for the next Recommendations has now been drafted and posted at the Commission's web site, www.icrp.org . The summary of that text is also included in the present issue.

Publication of the reports mentioned above in the open literature permitted rapid and extensive dissemination of these conceptual documents. However, this also meant that they were not specifically distributed to the *Annals of the ICRP* subscribers.

The Commission wishes to express its deep gratitude to the *Journal of Radiological Protection* and the Institute of Physics Publishing for allowing us to reproduce these copyrighted papers here, and to our own publishers, Elsevier Science Ltd., for accepting to publish the present issue as a supplement to Volume 34 of the *Annals of the ICRP*. This generosity allows us to provide a permanent record, at no cost to our subscribers, of the discussions leading up to the draft Recommendations that are now subject to public consultation.

During the period of preparation of the papers comprising this report, the membership of the Main Commission was:

(1997–2001)

R.H. Clarke (Chairman)	L.A. Ilyin	J.-C. Nénot
D. Beninson	A. Kaul	Z.Q. Pan
J.D. Boice jr	H. Matsudaira	B.C. Winkler
R. Cox	C.B. Meinhold (Vice-Chairman)	
L.-E. Holm	F.A. Mettler jr	

Scientific Secretary: J. Valentin

(2001–2005)

R.H. Clarke (Chairman)	A.J. Gonzlez	C. Streffer
R.M. Alexakhin	L.-E. Holm (Vice-Chairman)	A. Sugier
J.D. Boice jr	F.A. Mettler jr	Z.Q. Pan
R. Cox	R.J. Pentreath (2003-)	B.C. Winkler (✠ 2003)
G.J. Dicus	Y. Sasaki	

Scientific Secretary: J. Valentin

Emeritus members H.J. Dunster, B. Lindell, and W.K. Sinclair also participated in the October 1998 Main Commission meeting.

In April 1998, the Commission requested the Chairman to write the first of the three papers reproduced here under his own name. The two subsequent papers, written in the name of the Commission itself, were approved for publication in October 2000 and in January 2003, respectively. The draft Recommendations, the summary of which is reproduced here, was approved for release for public consultation in April 2004.

References

Clarke, R.H., 1999. Control of low-level radiation exposure: time for a change? J. Radiol. Prot. 19, 107–115.

ICRP, 1991. 1990 Recommendations of the International Commission on Radiological Protection. ICRP Publication 60, Ann. ICRP 21 (1–3).

ICRP, 2001. A report on progress towards new recommendations: a communication from the international commission on radiological protection. J. Radiol. Prot. 21, 113–123.

ICRP, 2003. The evolution of the system of radiological protection: the justification for new ICRP recommendations. J. Radiol. Prot. 23, 129–142.

SAGE

ICRP Supporting Guidance 4

Annals of the ICRP

Development of the Draft 2005 Recommendations of the ICRP: a Collection of Papers

ICRP Supporting Guidance 4

Commissioned/approved by ICRP, April 1998–April 2004

Abstract–In 1997, the International Commission on Radiological Protection (ICRP) initiated a project intended to lead up to the replacement of its 1990 Recommendations (*Publication 60*) with a view to producing new, consolidated Recommendations 10–15 years after those of 1990. In order to stimulate comprehensive discussion, an open and iterative consultation process was used. An initial set of conceptual proposals to be considered by the Commission, and two later progress reports describing how and why ideas evolved in interaction with the radiological protection community, were published in the *Journal of Radiological Protection* (*JRP*). These three papers are reproduced here with kind permission by the Institute of Physics Publishing. As a result of consultation on the initial conceptual proposals and subsequent debate, the Commission has now drafted a proposed text for its 2005 Recommendations. This is intended not as a radical revision, but as a more coherent statement of current policy and a simplification in its application. The summary of that draft is also reproduced here, while the full text, which is subject to public consultation until 31 December 2004, can be downloaded from the Commission's web site, www.icrp.org.

Keywords: Radiation protection, Constraints, Dose limits, Optimisation, Protection philosophy.

SAGE

ICRP Supporting Guidance 4

Annals of the ICRP

Development of the draft 2005 recommendations of the ICRP: a collection of papers – Part 2

Control of low-level radiation exposure: time for a change?

Roger Clarke

Chairman, International Commission on Radiological Protection, and
Director
National Radiological Protection Board, Chilton, Didcot, Oxon OX11 0RG, UK

Received 1 February 1999, in final form 26 February 1999, accepted for publication
1 March 1999

Abstract. The carcinogenic risks of exposure to low-level ionising radiation used by the
ICRP have been challenged as being, at the same time, both too high and too low. This
paper explains that the epidemiological evidence will always be limited at low doses, so
that understanding the cellular mechanisms of carcinogenesis is increasingly important to
assess the biological risks. An analysis is then given of the reasons why the challenges
to the ICRP, especially about the linear non-threshold response model, have arisen. As
a result of considering the issues, the Main Commission of the ICRP is now proposing a
revised, simpler, approach based on the concept of what is being called 'controllable dose'.
This is an individual-based philosophy and represents a shift in emphasis by the Commission
from societal-oriented criteria using Collective Dose. Finally the paper speculates on the
consequences for radiological protection of such a change in policy. The Commission
wishes its ideas to be discussed as part of its reconsideration of its recommendations.

1. Introduction

It is now ten years since the ICRP promulgated
a draft version of what was to become the
1990 recommendations. That consultation process
helped the Commission to clarify its aims and the
expression of its philosophy. Since the issue of
Publication 60 [1], the Commission has further
elaborated its policy on a number of issues such
as, control of exposure to radon-222, criteria for
intervention after an accident, the management
of occupational exposure, and its policy for the
disposal of radioactive wastes.

However, in recent years questions have been
raised about the Commission's application of its
risk factors at low doses. This article discusses
the current ICRP position and attempts to analyse
why the questions have arisen. Some proposals
are then made for a different, less complex,
approach to protection. The Commission is
considering a consolidation or recapitulation of its
1990 recommendations and wishes the ideas in this
paper to be widely discussed as part of the process
leading to a restatement of its protection policy.

2. Carcinogenic risks of low-level radiation exposure

2.1. Epidemiological evidence

Some of the most critical judgements in radiological protection have been associated with estimating the risk of excess cancer following low-
dose irradiation of human populations [1–4]. The
most difficult problem surrounding these judgements is that epidemiological approaches such as
those used with the Japanese A-bomb survivors
have only the power to identify excess risk down to
low-LET radiation doses of around 50–100 mGy
[5]. However, some analyses of the Japanese sur-
vivor data are claimed to show no excess below
200–300 mGy, and certainly some other cohorts
appear to demonstrate risks only at higher doses
than the data from the Japanese studies.

Below doses of a few hundred mGy, statistical power is progressively lost and direct estimates of cancer risk in a population of all ages becomes increasingly difficult and then impossible. Lower background cancer rates in children allow for estimation of *in utero* radiation risks down to about 10 mGy [3, 6], although these analyses are being challenged. But the problems of estimating the risk at occupational and environmental exposure levels of radiation remain.

Experimental limitations create essentially the same statistical problem in studies of animal carcinogenesis. However, in the last 10 years or so advances in biology, often based upon molecular genetics, have increasingly complemented the conclusions from epidemiology [2, 4].

2.2. Mechanisms of carcinogenesis

There is compelling evidence that cellular DNA present in the chromosomes of the cell nucleus acts as the principal target for spontaneously arising and carcinogen-induced tumours in humans and experimental animals [3, 4]. The DNA damage relevant to initial tumour development takes the form of gene and chromosome mutations that often appear to be specific to different tumour types.

There is abundant evidence that the capacity of irradiated cells to repair DNA damage acts to reduce mutational and tumorigenic risk. An argument used by some is that the low abundance of DNA damage at low doses allows complete and error-free cellular repair. According to these proposals it is only at high doses where repair capacity is saturated that tumorigenic risk becomes apparent. The proponents of this hypothesis support their argument with data showing that the abundance of spontaneously arising DNA damage arising in cells is very much greater than that induced by a low dose of ionising radiation, say 200 mGy—how can there be excess cancer risk at these low doses?

A large body of data reveals the critical flaw in this argument [4]. These data show clearly that spontaneously arising DNA damage is chemically simple, principally in single DNA strands and is readily repaired by the cell with a very low frequency of error, so that mutation rates are low. In contrast DNA damage produced by ionisation clusters within single radiation tracks is usually not chemically simple and can take the form of complex breaks in both strands of the DNA molecule. This complex damage is very difficult to repair correctly and as a consequence mutation rates are very much higher than that associated with spontaneous DNA damage. In accordance with these observations, dose-response relationships for gene and chromosomal mutations have been shown to be approximately linear down to doses of around 25 mGy, which is the statistical limit of their power. At present, the evidence available supports the view that ionising radiation acts most strongly as the early initiating phase of tumour development by inducing specific gene loss in stem cells [7].

Stated simply, although there are good reasons to believe that DNA damage repair in cells does act to substantially reduce the risk of radiation tumorigenesis, current knowledge does not support the concept that at low doses these repair functions can abolish such risk. Associated arguments for a dose threshold dependent upon the postulate that low-dose irradiation induces additional DNA repair capacity lack adequate supporting data and also fail to take account of the complex DNA damage problem noted above [8].

In the absence of directly informative quantitative data on radiation tumorigenesis, the shape of the low-dose response has to be judged on indirect data on the cellular mechanisms involved in the whole of this complex process.

In essence, this judgement has and will continue to be made on the basis of 'weight of evidence' since there are no prospects that the existence of a low-dose threshold for tumour induction could be proved or disproved conclusively. In respect of current knowledge it has been argued here that the evidence weighs against the concept of a low-dose threshold and favours the existing judgement that tumour risk will rise as a simple function of dose even at very low doses and dose rates. That is not to say that dose thresholds for tumour induction are not biologically feasible. Indeed data from experimental animals for certain tumour types and radiation quality do provide some evidence of this; one possible explanation of these data is that in some situations it is necessary to produce a degree of normal tissue damage before tumour development will proceed.

It is important to stress, however, that radiological protection systems need to be as simple as possible and to focus on the general consistency of all relevant data, not just the inevitable biological intricacies and exceptions.

The same general considerations apply to a controversy of more recent origin than that of threshold doses, namely the cellular phenomenon of radiation-induced persistent genomic instability [9]. It has been claimed by some [10] that the finding of this phenomenon poses a challenge to accepted concepts in radiological protection, and that risks may be higher than currently judged. The phenomenon has yet to be associated with tumour risk or other possible health effects [11]. Also, even if it were to be established, there would be no obvious implications for the direct epidemiological-based central estimates of cancer risk on which risk projections are founded. Nevertheless, the development of this new area of speculation on possible underestimation of low-dose risk provides an interesting counterpoint to the longer-standing debate on dose thresholds and the entirely opposite claims of its proponents.

In conclusion, ICRP judges that the weight of evidence at present falls in favour of assuming that those radiation events are potentially disruptive from the lowest doses. And while apoptosis, cellular surveillance, immune and adaptive responses are all real, they are most likely to modify the shape of the dose-response curve rather than proving a threshold [2,4].

The major policy implication of a non-threshold relationship for stochastic effects is that some finite risk must be accepted at any level of protection. Zero risk is not an option and this leads to the three principles that comprise the current policy of the Commission:

- Justification: do more good than harm.
- Optimisation: maximise the margin of good over harm.
- Limitation: Individual risk should not be unacceptable.

3. What is the problem?

It is useful to ask why it is that challenges to the so-called linear non-threshold hypothesis have arisen.

Contaminated land is an issue of considerable interest in many countries. It arises as a result of accidental releases, as from Chernobyl, and from man-made activities including atmospheric testing of nuclear weapons. Contamination is also an historic liability from, for example, luminising plants using radium, or from excessive effluent discharges.

A particular issue at present is the decommissioning of nuclear facilities, old reactors and weapons fabrication facilities. These liabilities require the expenditure of considerable amounts of money and some people think that too much money is being, and will be, spent to achieve low levels of residual contamination. If contaminated land is not cleaned up there is public concern and in some countries there will be litigation, charging that the environmental risk is too great. These concerns have led to an increased pressure from some individuals to propose a threshold in the dose-response relationship in order to reduce the expenditure. The issue is primarily in relation to public not occupational exposure.

Another aspect of concern is the use of Collective Dose to add up infinitesimally small doses to essentially infinite populations over essentially geological timescales and to cost it so that it is argued that it is worth committing huge resources today to protect the future. ICRP has already begun to tackle this by recommending, in Publication 77, the disaggregation of the single value of a collective dose into ranges of individual dose and the period of time when it is delivered. Further it cautions against the use of estimates of doses and health effects in the far future [12].

4. Difficulties with a threshold

A simple proportional relationship has important practical implications since it allows doses within an organ or tissue to be averaged over that organ or tissue, doses received at different times to be added, and doses from one source to be considered independently of the doses from other sources.

These practical implications are of overwhelming importance in radiological protection because of the complexity of the dose distributions in both space and time and because of the ubiquitous presence of natural sources of radiation. Very substantial difficulties would be introduced if threshold relationships were widely relevant in radiological protection. Threshold relationships exist for deterministic effects, but the levels of dose of concern in protection are generally well below these thresholds. When this is not so, as in radiotherapy, a single source of dose is predominant so that interaction between different sources can be neglected. One example of the complexities that would be introduced by a widely applicable threshold relationship would be the interaction between occupational exposure and non-occupational exposure to natural sources, and diagnostic medical exposure of individual workers. In order to control the risk it would be necessary to record all doses people received and with a threshold, protection by design is almost impossible. It is true that, increasingly, science is judged in the courts rather than by national academies of science. The judge and jury are increasingly likely to decide the issue and it is they who must be convinced as to whether there is a threshold and thus no risks at low doses of radiation.

As has been said above, there is uncertainty in risk estimates due to both biology and epidemiology, although it must be remembered that the exposures are always increments on the existing natural background radiation of a few mSv per year. Because of the continuing lack of definitive scientific evidence, a new approach to protection could be considered.

5. Confusion

ICRP has made clear that the present system of protection distinguishes between practices, which add doses and risks, and interventions, which reduce doses and risks [1, 12]. The dose limits apply to the sum of doses from a restricted set of sources or circumstances and, additionally, are often misunderstood, since a limit is sometimes taken to mean the boundary between safe and unsafe. For public exposure in particular, there is confusion about the application of the 1 mSv annual dose limit when the Action Level for radon in homes is to be set between 3 and 10 mSv in a year. Then, in the event of an accident, perhaps when people especially expect to be protected, the dose limit does not apply and intervention is not taken until doses are liable to be in the range of 5 to 50 mSv.

ICRP recommendations, in the context of the use of radionuclides, have been for the control of protection from single sources by optimisation within the individual maximum dose constraint of 0.3 mSv per year [12]. In the case of accidents, intervention levels have been suggested for taking action to reduce exposures, but there is no international guidance on the withdrawal of intervention actions. At what level of dose can normal living be resumed? More than 1 mSv per year surely, and if a new population moves from outside into the area, is it a practice to which the 1 mSv dose limit applies? Thus, at what point after an accident do the principles of protection for practices apply, if at all? Along these lines, is building a house in an area of high natural background radiation to which people might move from areas of lower background, a practice to which the 1 mSv limit is applied? Strict application of the definition of a practice given in ICRP Publication 60 might suggest that this is so.

These are situations that do not easily fall into the current definitions of practice or intervention; radiological protection philosophy might usefully be re-examined in order to develop an alternative logically consistent framework for protection to that used at present. The following thoughts are for discussion and are a first attempt to do this by bringing the three categories of exposure, occupational, medical and public, within an overall framework that encompasses the present system of protection for practices and interventions. These represent a scheme that may be complementary to, rather than a fundamental change in, the Commission's system of protection and may be of use in its application.

The difficulties outlined and the uncertainties in estimating risks from low-level radiation exposure have led ICRP to consider whether there might be some alternative way to deal with the control of dose. In formulating the proposals, an attempt has been made to try to simplify the system of protection.

6. A possible way forward

In protecting individuals from the harmful effects of ionising radiation, it is the control of radiation doses that is important, no matter what the source. Thus, a start may be made with a definition:

A *Controllable Dose* is the dose or the sum of the doses to an individual from a particular source that can reasonably be controlled by whatever means.

Such doses could be received at work, in medical practice and in the environment from the use of artificial sources of radionuclides, or could arise from elevated levels of natural radiation and radionuclides, including radon. The term covers doses that are being received, for example from radon, and doses that are to be received in the future, for example from the introduction of new sources or following an actual or potential accident. It does not apply to exposures that are not amenable to control, such as cosmic radiation at ground level, but would apply to high terrestrial levels of natural exposure.

In the past, ICRP has emphasized societal criteria, using collective dose summed over all populations and all times, principally in cost–benefit analysis, to determine the optimum spend on the control of a source. What is now being developed is a more individual-based philosophy, which was foreshadowed by the introduction of the concept of a constraint on the optimisation of a source and the Commission's recommendations on disaggregation regarding Collective Dose [12].

7. The principle

The protection philosophy for controllable dose is based on the individual. If the individual is sufficiently protected from a single source, then that is a sufficient criterion for the control of the source. The principle is

> If the risk of harm to the health of the most exposed individual is trivial, then the total risk is trivial—irrespective of how many people are exposed.

The significance of a level of controllable dose depends on its magnitude, the benefit to that individual and the ease of reducing or preventing the dose. There will, of course, be some level of dose where control will be mandatory. This will clearly be for the avoidance of deterministic effects in accident situations or for the protection of healthy tissues in high-dose medical procedures.

Doses of some hundreds of millisieverts up to several sieverts will cause deterministic effects of various types depending upon whether the exposure is acute or chronic. Apart from in radiotherapy, such doses may be encountered in interventional radiology, where there is a life-threatening situation. In other circumstances, such exposures will be entirely unacceptable to the individual, unless taken for life-saving rescue in an emergency. These situations are considered to be outside the scope of the proposed scheme of controllable doses set out here.

8. Controllable dose

For those exposures that are to be controlled, the philosophy is essentially set out here with a regime of controllable doses showing their different significance in terms of individual fatal cancer risk. In addition, the current criteria for controlling doses in normal, accident or medical situations are presented.

Thus, the highest dose that will normally be tolerated before control is definitely instituted is in the range of a few tens of millisieverts although this may be tolerated in successive years. This covers, *inter alia*:

- The permanent relocation of people following an accident is recommended to avert a lifetime dose of 1 Sv, which corresponds to some tens of mSv in the first year.
- The occupational dose limit of 20 mSv in a year.
- The upper (justified) action level for radon in homes (10 mSv per year).
- A CT scan (around 30–50 mSv).
- The lower level of averted dose above which evacuation is recommended after an accident (50 mSv).

The level of individual risk represented by some tens of mSv would be of the order of 1 in 1000 or 10^{-3}. While these levels of dose to the individual are not so high as to be completely unacceptable, they are levels at which questions should be asked as to whether the dose and associated fatal risk can be avoided by some sort of action. That action may be disruptive by intervening in lifestyle, or, as in the case of a CT scan, be simply to be sure that the required information cannot be obtained by another means, for example, magnetic resonance imaging.

Controllable doses should not generally exceed this level and actual or potential doses approaching this level would only be allowed if the individual receives a benefit or the doses cannot be reduced or prevented without significant disruption to lifestyle.

At levels of controllable dose of the order of a few millisieverts, the exposures should not be of great concern from the point of view of an individual's health. Natural background radiation is about 2–3 mSv in a year, and even if radon exposures are excluded, the figure is 1–2 mSv. Typical exposures in the range would be:

- The lower level of optimised range for radon intervention (3 mSv).
- The lower level for simple countermeasures (sheltering, KI) in an accident (5 mSv).
- The existing dose limit for members of the pubic (1 mSv).
- Simple diagnostic x-ray examinations (few mSv).

Steps may be taken to reduce these exposures, or to prevent them, particularly if the individual receives no benefit. Thus from a controllable dose of a few millisieverts upwards it becomes increasingly desirable to reduce or prevent the dose depending both on the practicability of doing so and whether the individual is deriving any tangible benefit from the exposure, for example annual occupational exposures or unnecessary doses from medical examinations. The associated levels of fatal risk would be 10^{-4}, 1 in 10 000.

In essence, this is a dose at which there is a question mark. If the medical examination is going to give a dose of a few mSv, again the question of whether an alternative procedure can give the required information should be asked, even though it can be argued that there is benefit to the patient. Similarly if a worker were receiving more than a few mSv, management would probably wish to ensure that the doses were as low as compatible with the job being undertaken. For the public, again action would be contemplated.

Doses that are below the millisievert level are also relevant in the control of exposures. In connection with uses of radiation sources, the Commission has set the maximum dose from a single source to a member of the public at 0.3 mSv a year [12]. The associated level of fatal cancer risk is about 10^{-5} per year. This level of dose is about 10% of total natural background dose and is also of the same order as to variation in background radiation (excluding the radon contribution) over much of the world. This level of imposed or involuntary risk is about the most that has been judged as being tolerated by members of the public.

In comparison, a level of risk of death of 10^{-6} per year is commonly regarded as trivial and the corresponding annual dose of about 10–20 μSv has been used to set exemption criteria for the Inter-Agency or European Basic Safety Standards [13, 14]. At this level of dose there should be no need to consider protection of the individual.

The dose levels discussed above are set out in figure 1, together with the doses that arise from the application of the present system of protection in a wide range of situations. There is, quite deliberately, no distinction between single doses and those that may be received repeatedly. This may be simpler for people to understand. Also, it is controversial to include medical exposures, but perhaps it may help to give the public a broader perspective on doses and risks if all the situations that lead to a given numerical value are put onto a single scale.

9. A practical solution

A suggested way forward may be to work toward a single maximum level of controllable dose. The value would be around 20–30 mSv in a year. Doses significantly above this level would only occur in uncontrolled accident situations or in life-saving medical procedures. It may be that rather than referring to this value as a limit, the term 'action level' should be used. In fact, that is what it would be—if controllable doses (actual or projected) are above this level action should be taken. This may have an advantage that Action Levels are understood, whereas a 'limit', as has been said, can be and often is misunderstood.

The management of controllable doses below the Action Level would be by individual-related source-specific Investigation Levels. They would apply to different actions taken to reduce exposures at the source, in the environment or by moving people. They would cover, for example,

CONTROLLABLE DOSE

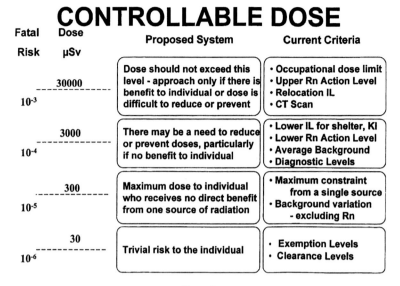

Figure 1.

occupational exposures, simple medical procedure doses, exposures from domestic radon or from other elevated levels of natural radionuclides, and those after an accident. The need for distinguishing between practices and interventions may no longer be required. This Investigation Level of around few millisieverts per year would prompt an investigation to see if anything simple could be done to reduce the exposure.

Within this scheme, exposures of a fraction of a millisievert would be the most that would ever be allowed to a member of the public from a single source, irrespective of the number of sources—effluents from a hospital, from a power plant, a diagnostic x-ray, a smoke detector, etc. These sources would be treated independently because the chance of one individual being exposed to all sources is very small and actual exposures from several sources would be unlikely to amount to more than a fraction of a millisievert. The term 'Constraint' could still be retained and the principle of optimisation applies for each source.

At the lowest level, doses of a few tens of microsieverts would be considered to be so low as to be beneath regulatory concern. There would

be no need to involve any system of protection below these levels.

10. The consequences

The proposals presented here put the primary emphasis for the system of protection on the individual, by adequately restricting the sources that may reasonably be controlled. The Commission's principles of justification and optimisation would next need to be reconsidered. Since radiological protection essentially plays such a minor part in a government's decision to justify the introduction, or the continuation, of a given use of radiation, consideration should be given to dropping the principle of justification from the ICRP system.

The existing principle of optimisation would be recast and clearly guidance would need to be developed on its application. This would require the replacement of 'as low as reasonably achievable', which has been associated with cost–benefit analysis and the use of Collective Dose, with another descriptor when individual dose is the determining criterion. It may be that the number of people affected by the highest levels of dose would be a determinant in deciding what is practicable.

The principles of protection might then become:

- Control the dose to the representative member of the most highly exposed group.
- Ensure that the resulting dose is 'as low as reasonably practicable'.

These may be known as 'Control' and 'ALARP'. There would be considerable scope for a simplification of the system of protection and remove confusion by not distinguishing between practices and interventions.

It is probably no longer sufficient for ICRP to state its belief that 'the standard of environmental control needed to protect man to the degree currently thought desirable will ensure that other species are not put at risk'. An advantage of the controllable dose system is that it may facilitate the development of an environmental protection strategy for radiation protection that is more compatible with those for other environmental agents.

Additionally, it may be that there is no longer a need to differentiate between occupational, public and medical exposures. The same guidance is equally applicable for protection of each category. Any particular concerns about the protection of the unborn child would also be covered, by the constraint of a fraction of a millisievert and investigation level of a few millisieverts.

There would be no need for the existing 1 mSv dose limit for the public.

Finally, there would be no use made of Collective Dose as currently defined, since the proposed policy of protection ensures that if the most exposed representative individual is sufficiently protected from a given source, then everyone else is also sufficiently protected from that source.

If at some time in the future it became possible that some individuals might be liable to receive, in due course and over a prolonged period of time, a significant accumulation of doses from many sources, local, regional and global, then a further restriction on sources may be necessary. There would, however, be likely to be a considerable time period available to effect change.

This more straightforward single-scale system of protection is consistent with the present system based on acceptable risks, but importantly may be explained to individuals more understandably as multiples or fractions of the natural background. In which case, perhaps there is no need to destroy the credibility of the profession in arguments for or against a threshold.

ICRP would welcome a wide discussion on the concepts of controllable dose and the new proposals for a simplification of protection philosophy that could lead to a restatement of its recommendations.

Résumé

On a contesté l'évaluation des risques carcinogéniques de l'exposition à de bas niveaux de rayonnement ionisant, employée par la CIPR; on la trouvait, soit trop élevée, soit trop faible. Dans cet article, on explique que l'évidence épidémiologique restera toujours restreinte, dans le cas des doses faibles; il en résulte que la compréhension des mécanismes cellulaires de la carcinogenèse est de plus en plus importante pour établir les risques biologiques. On analyse alors les raisons pour lesquelles il est apparu une récusation de la CIPR, en particulier en ce qui concerne le modèle de réponse linéaire sans seuil. Afin de sortir de cette situation, la commission principale de la CIPR propose maintenant un mode d'approche révisé, plus simple, fondé sur le concept de ce que l'on peut appeler la «dose contrôlable». Il s'agit d'une philosophie fondée sur l'individu; elle représente un déplacement d'accentuation par la commission, en partant des critères à orientation «sociétable» utilisant la dose collective. L'article s'achève par des spéculations quant aux conséquences d'un tel changement politique, en ce qui concerne la protection radiologique. La commission souhaite que ses idées soient discutées dans le cadre de la révision de ses recommandations.

Zusammenfassung

Die karzinogenen Risiken der Belastung durch von der ICRP eingesetzte schwachaktiven ionisierende Strahlung wurden in Zweifel gezogen, da sie sowohl zu hoch als auch zu niedrig seien. Diese Studie erklärt, daß die epidemiologischen Beweise bei niedrigen Dosen immer eingeschränkt sein werden, so daß ein Verständnis der zellulären Mechanismen der Karzinogenese immer wichtiger wird, um biologische Risiken abschätzen zu können. Es folgt dann eine Analyse der Gründe warum Zweifel an der ICRP, insbesondere am linearen Ansprechmodell ohne Schwellen, aufgekommen sind. Als Ergebnis der Überlegungen zu diesen Fragen schlägt die Hauptkommission des ICRP nun eine überarbeitete, einfachere Methode vor, basierend auf dem Konzept der sogenannten 'kontrollierbaren Dosis'. Dies ist eine Personen-bezogene Philosophie, die einen Verschiebung der Gewichtung durch die Kommission von gesellschaftlich orientierten Kriterien unter Verwendung einer kollektiven Dosis widerspiegelt. Schließlich spekuliert die Studie über die Konsequenzen einer derartigen Veränderung der Politik für den Strahlenschutz. Die Kommission möchte, daß ihre Ideen als Teil der Überprüfung ihrer Empfehlungen diskutiert werden.

References

[1] International Commission on Radiological Protection (ICRP) 1990 Recommendations of the International Commission on Radiological Protection *ICRP Publication* 60 (*Ann. ICRP* **21** (1-3) 1991

[2] UNSCEAR 1993 Sources and effects of ionizing radiation *1993 Report to the General Assembly with Annexes* (New York: United Nations)

[3] Muirhead C R, Cox R, Stather J W, MacGibbon B H, Edwards A A and Haylock R G E 1993 Estimates of late radiation risks to the UK population *Doc. NRPB* **4** (4)

[4] Cox R, Muirhead C R, Stather J W, Edwards A A and Little M P 1995 Risk of radiation-induced cancer at low doses and low dose rates for radiation protection purposes *Doc. NRPB* **6** (1)

[5] Pierce D A, Shimizu Y, Preston D L, Vaeth M and Mabuchi K 1996 Studies of the mortality of atomic bomb survivors. Report 12, part 1, cancer 1950–1990 *Radiat. Res.* **146** 1–27

[6] Doll R and Wakeford R 1997 Risk of childhood cancer from fetal irradiation *Br. J. Radiol.* **70** 130–9

[7] Cox R 1997 The mechanisms and genetics of radiation tumorigenesis: recent developments in animal models *Health Effects of Low Dose Radiation* (London: British Nuclear Energy Society) pp 21–4

[8] UNSCEAR 1994 Sources and effects of ionizing radiation *1994 Report to the General Assembly, with Scientific Annexes* (New York: United Nations)

[9] Morgan W F, Day J P, Kaplan M I, McGee E M and Limoli C L 1996 Genomic instability induced by ionising radiation *Radiat. Res.* **146** 257–8

[10] Edwards R 1997 Radiation roulette *New Sci.* 11 October 37–40

[11] Stather J W and Cox R 1998 Radiation risks and genomic instability (correspondence) *EULEP, EURADOS and UIR Newsletter* (3) 5–7

[12] International Commission on Radiological Protection (ICRP) 1978 Radiological protection policy for the disposal of radioactive wastes *ICRP Publication* 77 (*Ann. ICRP* **27** (suppl)

[13] IAEA 1996 International basic safety standards for protection against ionizing radiation and for the safety of radiation sources *IAEA Safety Series* 115 (Vienna: IAEA)

[14] Council of the European Union 1996 Council directive 96/29/Euratom of 13 May 1996 laying down basic safety standards for the protection of the health of workers and the general public against the dangers arising from ionizing radiation *Official Journal of the European Communities* **39** L159 (29 June 1996)

INSTITUTE OF PHYSICS PUBLISHING   JOURNAL OF RADIOLOGICAL PROTECTION

J. Radiol. Prot. **21** (2001) 113–123   www.iop.org/Journals/jr   PII: S0952-4746(01)22869-9

MEMORANDUM

A report on progress towards new recommendations: A communication from the International Commission on Radiological Protection

International Commission on Radiological Protection[1]

ICRP, SE-171 16, Stockholm, Sweden

Received 2 March 2001

Abstract

Throughout the hundred-year history of the uses of ionising radiation in medicine and industry there has been advice on the need to protect people from the hazards associated with exposure. Protection standards have evolved throughout this period to reflect both the scientific understanding of the biological effects of exposure and the social and ethical standards to be applied. The Main Commission of ICRP is now considering a revised, simpler approach that is based on an individual-oriented philosophy and represents a potential shift by the Commission from the past emphasis on societal-oriented criteria. The initial proposals were promulgated through IRPA and an open literature publication was published in the *Journal of Radiological Protection* in June 1999. On the basis of comments received and the observations presented at the IRPA-10 Congress in May 2000, the Commission is beginning to develop the next recommendations. This article describes the issues involved in the preparation of the next recommendations and indicates the process that the Commission proposes to follow. The Commission wishes there to be an ongoing debate with an iteration of ideas over the next few years.

1. Historical background

Roentgen discovered x-rays in 1895, and in 1896 Grubbé described x-ray dermatitis of hands in the first paper to appear reporting radiation damage to the skin of the hands and fingers of the early experimental investigators. On the 12 December 1896, the American journal *Western Electrician* contained a paper by Wolfram Fuchs giving the first protection advice. This was:

- make the exposure as short as possible;
- do not stand within 12 inches (30 cm) of the x-ray tube; and
- coat the skin with Vaseline and leave an extra layer on the area most exposed.

Becquerel's identification of the phenomenon of radioactivity, also in 1896, and the Curies' separation of radium in 1898 soon led to the use of radioactive substances, together with x-rays, for therapy. In the next ten years, many papers were published on the tissue damage caused by radiation.

[1] Presented by the ICRP Chairman, Roger H Clarke. E-mail: roger.clarke@nrpb.org.uk

In 1913, the Deutsche Roentgen Gesellschaft issued radiological protection advice and in 1915 the British Roentgen Society recognised the hazards of x-rays in a warning statement. Several countries were actively reviewing standards for safety by the start of the First World War, but it was not until 1925 that the International Congress of Radiology was formed and first met to consider establishing protection standards. This Congress established the 'International X-ray and Radium Protection Committee' in 1928, which evolved into the present International Commission on Radiological Protection (ICRP) in 1950.

The early recommendations were concerned with avoiding threshold (*deterministic*) effects, initially in a qualitative manner. A system of measurement or dosimetry was needed before protection could be quantified and dose limits could be defined. In 1934, recommendations were made implying the concept of a safe threshold (ICRP 1934): 'Under satisfactory working conditions a person in normal health can tolerate exposure to x-rays to an extent of about 0.2 roentgens per day.' This would be about ten times the present annual dose limit. The tolerance idea continued for the next two decades so that in the 1950 recommendations (ICRP 1951) it is stated that 'the figure of 2 r per week seems very close to the probable threshold for adverse effects'. This led to a proposed limit of 0.3 r per week for low-LET radiation. In considering neutrons and alpha particles, it was stated that 'anaemia and bone damage appear to have a threshold at 1 μCi Ra-226'. In these 1950 recommendations, the Commission provided an impressive list of the health effects that should be kept under review:

- Superficial injuries.
- General effects on the body, particularly blood and blood-forming organs, e.g. production of anaemia and leukaemia.
- The induction of malignant tumours.
- Other deleterious effects including cataract (and other less likely examples).
- Genetic effects.

For the first 60 years after the discovery of ionising radiation, the ethical position was that of avoiding deterministic effects from occupational exposures and the principle of radiological protection was to keep *individual doses* below the relevant *thresholds*. Low doses of radiation were deemed beneficial, largely because the uses of radiation were for medical purposes, and radioactive consumer products abounded.

A change in philosophy was brought about by new biological information that began to emerge in the mid-1950s. There was the epidemiological evidence of excess malignancies amongst American radiologists and the first indication of an excess of leukaemia cases in the survivors of the atomic bombings at Hiroshima and Nagasaki. Previously there had been only deterministic effects, where the severity of the effect increases with the size of the dose, and above a certain threshold dose the effect is almost certain to appear. Now there were *stochastic effects* where the probability of the effect, not the severity, is proportional to the size of the dose.

The threshold was rejected. The problem had become one of limiting the probability of harm, and much of what has subsequently developed related to the estimation of that probability of harm and the decision on what level of implied risk is acceptable or, more importantly, unacceptable. In the 1955 recommendations, ICRP first began to address this question of acceptability (ICRP 1955). It was said that since no radiation level higher than the natural background level can be regarded as absolutely safe, the problem is to choose a practical level that, in the light of present knowledge, involves a negligible risk. Maximum permissible doses should be set so as to involve a risk which is small compared with other hazards in life. And: 'In view of the incomplete evidence on which the (risk) values are based coupled with the knowledge that some effects are irreversible and cumulative ... it is strongly recommended that every effort be made to reduce exposure to all types of ionising radiation to the lowest possible level.'

There was then a prolonged debate over how to deal with the acceptability of the risks. In Publication 1 (ICRP 1959), the words 'lowest possible' were succeeded by 'as low as practicable' and by 1966 had become 'as low as is readily achievable' (ICRP 1966). The Commission used these words so as to include social and economic considerations. Other considerations, such as ethical ones, were not excluded by this wording; the Commission considered them included in the adjective 'social'. In Publication 22 (ICRP 1973), the adverb 'readily' was replaced by 'reasonably', because 'readily' seemed to be too permissive.

The 1977 recommendations (Publication 26) (ICRP 1977) set out the new system of dose limitation and introduced the three principles of protection in paragraph 12:

- No practice shall be adopted unless its introduction produces a positive net benefit.
- All exposures shall be kept as low as reasonably achievable [ALARA], economic and social factors being taken into account.
- The doses to individuals shall not exceed the limits recommended for the appropriate circumstances by the Commission.

These principles have since become known as *Justification of a Practice*, *Optimisation of Protection (or ALARA)* and *Individual Limits*.

The recommendations were much concerned with the bases for deciding what is reasonably achievable in dose reduction. The principles of justification and optimisation aim at doing more good than harm and at maximising the margin of good over harm for society as a whole. They therefore satisfy the *utilitarian principle of ethics*, whereby actions are judged by their overall consequences, usually by comparing in monetary terms the relevant benefits (e.g. statistical estimates of lives saved) obtained by a particular protective measure with the net cost of introducing that measure.

Paragraph 72 of Publication 26 suggests that the decision on what is ALARA depends on the answer to the question: is the collective dose sufficiently low that further reduction in dose would not justify the incremental cost required to accomplish it? Paragraph 75 then recommended the use of differential cost–benefit analysis where the independent variable is the collective dose, and recommended that there be assigned a monetary value to a unit of collective dose. This classical use of cost–benefit analysis addresses the question: how much does it cost and how many lives are saved?

In 1977, the establishment of the dose limits was of secondary concern to the cost–benefit analysis and use of collective dose. This can be seen in the wording used by ICRP in setting its dose limit for members of the public. Publication 26 states: 'The assumption of a total risk of the order of 10^{-2} Sv^{-1} would imply restriction of the lifetime dose to the individual member of the public to 1 mSv per year. The Commission's recommended limit of 5 mSv in a year, as applied to critical groups, has been found to give this degree of safety and the Commission recommends its continued use.' In a similar manner the dose limit for workers was argued on a comparison of average doses and therefore risk in the workforce with average risks in industries that would be recognised as being 'safe', and not on maximum risks to be accepted.

Throughout this period of protection, the Commission was dealing with stochastic risks where the probability of harm was proportional to dose. The question had become one of acceptability of risk, since there was no threshold below which there was zero risk. This acceptability was determined by what was 'as low as reasonably achievable' and the utilitarian ethical approach was used. This approach provided insufficient protection for the individual and required the retention of the concept of the dose limit.

In 1989, the Commission issued the draft of a new set of recommendations to be issued in 1990. Several developments had led the Commission to revise its recommendations. The most powerful argument was the availability of new information suggesting higher values of the probability of stochastic effects of radiation. The dose limit for public exposure applied only in defined conditions, but many people regarded a limit as being absolute. The use of higher doses for emergencies and for radon in homes was seriously confusing. The Commission tried to clarify this by distinguishing between *practices* that added doses and *interventions* that subtracted doses in existing situations. The confusion was not eliminated. Other factors included the excessive formality of the use of differential cost–benefit analysis and the rigid interpretation of collective dose. A report on the specific topic of optimisation, Publication 55 (ICRP 1989), was issued to reduce this formality, but did not seem to have much influence.

2. The present situation

The 1990 recommendations were issued as Publication 60 (ICRP 1991a). They still adopted the same three principles, in the same order, but extended the explanations of the 1977 material into a 'system of radiological protection':

- No practice involving exposures to radiation should be adopted unless it produces sufficient benefit to the exposed individuals or to society to offset the radiation detriment it causes.
- In relation to any particular source within a practice, the magnitude of individual doses, the number of people exposed, and the likelihood of incurring exposures where these are not certain to be received should all be kept as low as reasonably achievable, economic and social factors being taken into account. This procedure should be *constrained by restrictions on the doses to individuals* (dose constraints), or on the risks to individuals in the case of potential exposures (risk constraints) so as to limit the inequity likely to result from the inherent economic and social judgements.
- The exposure of individuals resulting from the combination of all the relevant practices should be subject to dose limits, or to some control of risk in the case of potential exposures. These are aimed at ensuring that no individual is exposed to radiation risks that are judged to be unacceptable from these practices in any normal circumstances.

In the 1990 recommendations, the Commission continued to adopt implicitly a societal/ethical policy using a *utility-based criterion*, aimed at determining the optimum deployment of resources applied to optimise protection by the control of, or at, a source. Because of the emphasis on collective dose in the first two requirements, this order emphasised the protection of society over that of individuals. However, this emphasis does not necessarily provide sufficient protection for each individual. Classical cost–benefit analysis is unable to take this into account, so the Commission established an added restriction on the optimisation process. This addition modified the principle of optimisation by the introduction of the concept of a constraint. Optimisation is a source-related process while limits apply to the individual to ensure protection from all sources under control. The constraint is an *individual-related criterion*, applied to a single source in order to ensure that the most exposed individuals are not subjected to excessive risk, and to limit the inequity often introduced by cost–benefit analysis.

The definition of the dose limits was changed to indicate that the continued exposure just above the limits would result in additional risks that could reasonably be described as 'unacceptable in normal circumstances'. Much of Publication 60 was concerned with how the level of unacceptability should be established. It therefore included comprehensive annexes on dosimetric quantities, biological effects, and the bases for judging the significance of biological effects. It has however proved complex and, in parts, confusing. For example:

- The users of the recommendations confused justification and the optimisation of protection.
- The optimisation requirement had led to an overemphasis of the use of differential cost–benefit analysis and collective dose, thereby losing the emphasis on 'reasonably' in the phrase 'as low as reasonably achievable'.
- The use of collective dose, aggregated to include all levels of dose and all periods of time into a single value, distorted the process of optimisation of protection.
- The dosimetric quantities were not directly measurable and caused concern in relation to measurable quantities.

The Commission weakened the link to cost–benefit analysis and collective dose, initially in Publication 55 (ICRP 1989) and more firmly in Publication 77 (ICRP 1997), reflecting an overall shift of the ethical position from utilitarian values. However, the Commission concluded that it should begin the process to produce a new set of recommendations, at a date of about 2005. This will be 15 years after the adoption of Publication 60.

The Chairman of the Main Commission made proposals to the Commission for a possible simplification of the system of protection, which were accepted. The principal change was to emphasise the dose to an individual from a controllable source. There would still be requirements to keep the individual dose both below a defined action level and as low as reasonably practicable. The second requirement would not be linked to collective dose in its present form. The Commission invited its Chairman to publish the proposals (Clarke 1999) and the International Radiation Protection Association (IRPA) to arrange a discussion of the Chairman's paper amongst Constituent Societies and to make comments.

IRPA held a special session at its Congress, IRPA-10, in May 2000 where the principal views expressed can be summarised as follows. The majority of Societies indicated the following views:

- Essentially all Societies welcomed the Commission's initiative to open the debate.
- There was an overwhelming view that the principle of Justification must be addressed.
- The majority of Societies agreed with the use of natural background radiation in setting and explaining protection standards.
- The retention of limits was felt essential.
- There was thought to be some value in the use of collective dose for workers.

A minority of Societies argued for other views:

- There was a need for the retention of the unrestricted (infinite time, infinite space) collective dose.
- The system was not applicable in medical areas.
- Changes should be made only if there would be an improvement in protection.

The Commission has considered these opinions and has decided to proceed, taking account of views expressed, towards new recommendations. It wishes there to be an on-going debate with an iteration of ideas in the development process. This article describes the principal issues involved in the preparation of the next recommendations.

3. The development of the next recommendations

A great deal of work has still to be undertaken by the Commission and its Committees. A Task Group of the Main Commission has already been selected to co-ordinate the programme. The Committees will develop proposals for discussion by the Task Group which will develop position papers on the many issues to be resolved before the Commission finalises any new recommendations. The material in this communication is therefore intended only to indicate the areas in which the Task Group will be expected to work.

3.1. The system of protection

It is the control of radiation doses that is important, no matter what the source, in protecting individuals from the harmful effects of ionising radiation. In most situations, the most effective controls are those applied at or near the source of radiation. In the first place, therefore, consideration should be given to the dose to an individual from a particular controllable source.

The term *controllable source* can be used when either the source or the resulting exposures are controllable by reasonable means. It may be more convenient to call a controllable source an *optional source* when the existence, or the nature, of the source is a matter of choice. The term *unavoidable source* might be used when neither the existence nor the nature of the source is a matter of choice, but the pathways to man are controllable.

The doses may be received as result of work (occupational exposure), in medical practice (medical exposure), in the environment (public exposure), due to artificial radionuclides, or to natural sources such as cosmic rays and long-lived radionuclides in the Earth's crust. The doses may have already been received, or will be received in the future, from the introduction of new sources or following an actual or potential accident.

For each previously justified, controllable source, the first consideration in the proposed system of protection is to restrict the dose to individuals by means of *Protective Action Levels*. The need for protective action is influenced by the individual dose, but not by the number of exposed individuals. At present, this criterion is provided by dose limits, constraints and action, or intervention, levels. The second consideration stems from the recognition that there is likely to be some risk to health, even at small doses. This introduces a moral requirement, for each controllable source, to take all reasonable steps to restrict both the individual doses to levels below the action level and the number of exposed individuals. At present, the Optimisation of Protection provides this criterion.

3.2. The justification of an endeavour

'Endeavour' might be a better description than the current term 'practice' which has proved difficult, partly because of its antithesis, 'theory'. The judgement that it would be justifiable to introduce or continue an endeavour involving exposure to ionising radiation is important, but is not usually taken by radiological protection authorities, although they should influence the decision. The responsibility for judging the justification of an endeavour usually falls on governments or government agencies.

The medical exposure of patients in a general sense should be justified, as is any other endeavour. In addition, a more detailed justification has to be introduced. The principal aim of medical exposures is to do more good than harm *to the patient*, subsidiary account being taken of the radiation detriment from the exposure of the radiological staff or of other patients. Provided that the necessary resources are available, the responsibility for the justification of the particular use of a particular procedure falls on the relevant medical practitioners.

The Commission's present recommendations for justification require that the endeavour should do more good than harm. This procedure implies a quantified balance of costs and benefits, but in practice, governments, physicians or individuals do not make decisions about courses of action in a predominantly quantitative way. A qualitative approach is more common and usually more appropriate. Nevertheless, a strict interpretation of the present system has

Table 1. Bands of concern about individual effective doses in a year.

Band of Concern	Description	Level of dose
Band 6	Serious	> 100 × normal
Band 5	High	> 10 × normal
Band 4	Normal	1–10 mSv
		(Typical natural background)
Band 3	Low	> 0.1 × normal
Band 2	Trivial	> 0.01 × normal
Band 1	Negligible	< 0.01 × normal

often been taken to require an estimate of the total collective dose from the endeavour, over all levels of dose, all locations and all times.

The next recommendations will apply only to situations in which either the source is susceptible to control (by elimination or modification), or the pathways to individual dose are subject to control, or both. Furthermore, the current proposals for the new system of protection start from the justification of an endeavour. Until justification has been established in general terms, it is not considered appropriate to apply a system of protection to optional sources. The next recommendations as now proposed would therefore apply only to justified optional sources and to unavoidable sources.

3.3. Protective Action Levels

The proposed system of protection starts from a generalised structure of individual doses linked to recommended Protective Action Levels. These are levels of individual dose above which it is proposed that there is a requirement to take all feasible steps to reduce doses. They are influenced by the type of action and by the type of exposed individual. This necessitates a number of such levels. Protective Action Levels would be chosen on the assumption that the action is, or will be, effective.

A convenient starting point for the use of Protective Action Levels is a classification of levels of individual dose. A scale indicating the appropriate level of concern was suggested by the Commission (Clarke 1999). The aim was to specify a broad basis for defining bands of concern. It is desirable to avoid a rigid demarcation of the bands while avoiding ambiguity. Table 1 provides a similar classification into bands, each with a descriptive specification, and an indication of the level of dose compared with natural background.

The most effective action will be that applied at the design stage of a new endeavour. However, experience may show that the design precautions were inadequate, or accidents may occur. Existing precautions may sometimes have to be improved. If the source is not optional, additional protective action may have to be applied in the environmental pathways, or to individuals, previously known as intervention. Protective actions will be different for optional sources and unavoidable sources.

The bands of concern suggested in table 1 provide guidance about the ranges of action level relevant to common types of protective action and types of exposed individuals. Broad principles may be included in the next recommendations. Detailed guidance will probably be best provided in specialist publications of the Commission.

The magnitude of Protective Action Levels and the methods of application will have to be decided by the Commission at a later stage of the preparation of the next recommendations. Meanwhile, table 2 uses the bands of concern given in table 1 to provide guidance about the broad range of action required to meet relevant protective action levels. At exposures below the

Table 2. Typical protective actions for optional and unavoidable sources.

Level of concern	Type of exposure	Typical protective actions for optional sources	Typical protective actions for unavoidable sources
Band 6 (Serious)	Public	Remove or greatly reduce the source	Relocate or temporarily evacuate individuals
	Medical	Justify the exposure (excluding therapy)	Assess the consequences, treat the individuals if necessary
	Occupational	Remove or greatly reduce the source	Assess the consequences, treat the individuals if necessary
Band 5 (High)	Public	Reduce the source	Shelter in buildings. Administer stable iodine
	Medical (diagnosis)	Reconsider the diagnostic procedure	Assess the implications
	Occupational	Reduce the dose	Reduce the dose
Band 4 (Normal)	Public	Reduce the dose	Reduce the dose
	Medical (diagnosis)	Reconsider the diagnostic procedure	No protective action
	Occupational	Reconsider the working procedure	No protective action
Band 3 (Low)	Public	Reduce the dose	No protective action
	Medical (diagnosis)	No protective action	No protective action
	Occupational	No protective action	No protective action
Band 2 (Trivial)	Public	No protective action	No protective action
	Medical (diagnosis)	No protective action	No protective action
	Occupational	No protective action	No protective action
Band 1 (Negligible)	Any exposure	Exclude from the ICRP system of protection	Exclude from the ICRP system of protection

action level, there would be a necessary, but less prescriptive, requirement to take all reasonable steps to achieve further reductions in doses. This is part of the process to keep exposures as low as is reasonably practicable (ALARP).

3.4. Optimisation of Protection

The process of taking all reasonable action to reduce exposures is still likely to be called the Optimisation of Protection. The initial proposals (Clarke 1999) suggested that the optimisation of protection as it is now usually understood should be replaced by a different requirement to ensure that the residual doses, after the application of the Protective Action Levels, should be kept 'as low as reasonably practicable' (ALARP). This requirement would apply both to individuals and to groups.

One procedure for judging that the doses are as low as is reasonably practicable would involve the comparison of a number of feasible protection plans. The comparison would aim at selecting the plan where the step to the plan next higher in stringency would result in an improvement insufficient to offset the increase in resources needed to take the step. The current plan could then be said to result in exposures that are as low as reasonably practicable. Since the requirement would apply to both individuals and groups, the choice would be dependent on judgement rather than on collective dose.

In addition to the protection of individuals, there is an additional need to protect groups, taking account of changes both in the level of dose and the number of individuals affected. The obvious quantity to use is the product of the size of the exposed group and the average dose to the group. The term collective dose has proved unsuitable for this product because it is widely used for the product over the world population and over all time. The Commission is considering the term *group dose*, where the definition of the exposed group is limited to individuals receiving specified ranges of dose over specified periods of time. The use of the group dose would be particularly useful when the group is a workforce. It might then be called the *workforce dose*.

The requirement to keep all doses as low as is reasonably practicable should also take account of the possible inequity of the distribution of individual doses. This inequity will be limited by the use of Protective Action Levels, but the formal use of *reference levels* will still be needed to express optimised protection. These have already been used for medical diagnostic procedures in the Euratom Patient Protection Directive of the European Community and also in the IAEA Basic Safety Standards as guidance levels.

4. Other issues

4.1. Health effects

In the 1990 recommendations (ICRP 1991a), the Commission included the following material: 'The probability of a cancer resulting from radiation usually increases with increments of dose, probably with no threshold, and in a way that is roughly proportional to dose, at least for doses well below the thresholds for deterministic effects. The severity of the cancer is not affected by the dose. This kind of effect is called 'stochastic', meaning 'of a random or statistical nature'. If the damage occurs in a cell whose function is to transmit genetic information to later generations, any resulting effects, which may be of many different kinds and severity, are expressed in the progeny of the exposed person. This type of stochastic effect is called 'hereditary'.'

The systems of protection in the 1990 recommendations and suggested in this article are based on these views. The Commission has initiated a comprehensive review of the biological and epidemiological information. If this review leads to a major change in these views, the Commission will have to judge whether the system now suggested, which will be effective and prudent, can be maintained. The introduction of a threshold for stochastic effects or a serious departure from linearity would be inconsistent with the use of average absorbed dose and with the independence of the assessment of the effects of individual sources. The whole conceptual basis of radiological protection would be changed. For the present, no such conceptual change is considered.

4.2. Dosimetric quantities

There have been some persistent differences of view about the definitions of the Commission's dosimetric quantities. In the next recommendations, the Commission hopes to remove these differences. The root of the conflict lies in the difference in objectives. Metrologists have correctly aimed at providing a definition of 'dose' that unequivocally defines the dose in an absorbing medium, without involving in the definition itself information about the type of radiation or the size, shape and composition of the medium.

This aim has led to the quantity *absorbed dose at a point*. Whatever the exposure situation, the absorbed dose can, in principle, be determined using only information available at the point. For protection purposes, the aim is different. The dosimetric quantity should correlate reasonably well with the probability and severity of the consequent health effects. Clearly, this aim can never be fully achieved, and fairly crude relationships will always have to be accepted. Absorbed dose at a point is a good starting point, but the information at a point is unlikely to meet the protection aims. For protection purposes, the Commission uses the *average tissue dose*, sometimes weighted for the type of radiation and for the choice of tissue. The magnitude of the average tissue dose depends on the type of radiation and on the shape, size, composition and location of the tissue. None of this information is included in the definition, so it is always necessary to state, or to imply, the details of the models used. In the next recommendations, it will be necessary to clarify the difference between the quantities that can be specified at a point and the quantities that are averaged over tissue.

The radiation and tissue weighting factors are derived from biological data, but each numerical value draws on a wide range of situations. The values cannot be taken directly from the experimental data derived from particular studies. The sets of values currently in use do not reflect this generalisation and are more complex than can be justified. Consideration will be given to providing simpler sets of weighting factors in the next recommendations. The aims of the weighting factors are not likely to change and it may be possible to define equivalent dose and effective dose in terms that will not cause quantitative changes large enough to be significant for radiological protection purposes. In any event, it should be possible to separate the measurable quantities, which are subject to conventional metrology, and the protection quantities, which would be determined from the measurable quantities by conversion coefficients chosen by ICRP.

4.3. Medical applications

In Publication 73, 'Radiological Protection and Safety in Medicine' (ICRP 1996), the Commission explained the features of medical practice that influence the approach to radiological protection. The following paragraph is taken from Publication 73:

> Several features of medical practice require an approach to radiological protection that is slightly different from that in other practices. In the first place, the exposure of patients is deliberate. Except in radiotherapy, it is not the aim to deliver a dose of radiation, but rather to use the radiation to provide diagnostic information or to conduct interventional radiology. Nevertheless, the dose is given deliberately and cannot be reduced indefinitely without prejudicing the intended outcome. Secondly, the patient needs a special relationship with the medical and nursing staff. For this reason, the system of protecting the staff from the source, e.g. shielding, should be designed to minimise any sense of isolation experienced by the patient. This is particularly relevant in nuclear medicine and brachytherapy, where the source is within the patient. Thirdly, in radiotherapy, the aim is to destroy the target tissue. Some deterministic damage to surrounding tissue and some risk of stochastic effects in remote non-target tissues are inevitable. Finally, hospitals and radiology facilities have to be reasonably accessible to the public, whose exposure is thus more difficult to control than it is in industrial premises.

Biomedical research includes the exposure of volunteers who do not necessarily obtain any benefit from the exposure. The associated problems were discussed in Publication 62, 'Radiological Protection in Biomedical Research' (ICRP 1991b). The Commission intends to retain these approaches in the next recommendations.

5. Conclusions

The discussions in this article indicate how it will be feasible to develop the next recommendations based on an individual-related philosophy using the concept of controllability of sources. For each previously justified, controllable source the first consideration in the proposed system of protection would be to restrict the dose to individuals by means of Protective Action Levels. There still remains an additional requirement to do all that can be done to make exposures as low as reasonably practicable (ALARP) from the source. Thus a major change from Publication 60 will be this re-ordering of the principles, to place that of optimisation after that of individual protective action. This is a development signalled in Publication 60 by the introduction of a constraint to the optimisation of protection.

A great deal of work has still to be undertaken by the Commission and its Committees. A Task Group of the Main Commission has already been selected to co-ordinate the programme and the Committees will be asked to develop position papers on the many issues to be resolved before the Commission finalises any new recommendations.

References

Clarke R 1999 Control of low-level radiation exposure: time for a change? *J. Radiol. Prot.* **19** 107–15

ICRP 1934 International recommendations for x-ray and radium protection *Br. J. Radiol.* **7** 1–5

ICRP 1951 International recommendations on radiological protection *Br. J. Radiol.* **24** 46–53

ICRP 1955 Recommendations of the International Commission on Radiological Protection *Br. J. Radiol.* (suppl 6)

ICRP 1959 *Recommendations of the International Commission on Radiological Protection (Publication 1)* (Oxford: Pergamon)

ICRP 1966 *Recommendations of the International Commission on Radiological Protection (Publication 9)* (Oxford: Pergamon)

ICRP 1973 *Implications of Commission Recommendations that Doses be Kept as Low as Readily Achievable (Publication 22)* (Oxford: Pergamon)

ICRP 1977 Recommendations of the International Commission on Radiological Protection (Publication 26) *Ann. ICRP* **1** (3)

ICRP 1989 Optimisation and Decision-Making in Radiological Protection (Publication 55) *Ann. ICRP* **20** (1)

ICRP 1991a 1990 Recommendations of the International Commission on Radiological Protection (Publication 60) *Ann. ICRP* **21** (1–3)

ICRP 1991b Radiological Protection in Biomedical Research (Publication 62) *Ann. ICRP* **22** (3)

ICRP 1996 Radiological Protection and Safety in Medicine (Publication 73) *Ann. ICRP* **26** (2)

ICRP 1997 Radiological Protection Policy for the Disposal of Radioactive Waste (Publication 77) *Ann. ICRP* **27** (suppl)

MEMORANDUM

The evolution of the system of radiological protection: the justification for new ICRP recommendations

The International Commission on Radiological Protection[1]

ICRP, SE-171 16, Stockholm, Sweden

Received 18 February 2003
Published 9 April 2003
Online at stacks.iop.org/JRP/23/129

Abstract

ICRP has been encouraging discussion during the past few years on the best way of expressing radiological protection philosophy in its next recommendations, which it plans to publish in 2005. The present recommendations were initiated by Publication 60 in 1990 and have been complemented by additional publications over the last 12 years. It is now clear that there is a need for the Commission to summarise the totality of the number of numerical values that it has recommended in some ten reports. This has been done in this paper, and from these a way forward is indicated to produce a simplified and more coherent statement of protection philosophy for the start of the 21st century. A radical revision is not envisaged, rather a coherent statement of current policy and a simplification in its application.

1. Introduction

The 1990 system of protection, set out in Publication 60 (ICRP 1991), was developed over some 30 years. During this period, the system became increasingly complex as the Commission sought to reflect the many situations to which the system applied. This complexity involved the justification of a practice, the optimisation of protection, including the use of dose constraints, and the use of individual dose limits. It has also been necessary to deal separately with endeavours prospectively involving radiation exposure, '*practices*', for which unrestricted planning was feasible for reducing the expected increase in doses, and existing situations for which the only feasible protection action was some kind of '*intervention*' to reduce the doses. The Commission also considered it necessary to apply the recommendations in different ways to occupational, medical and public exposures. This complexity is logical, but it has not always been easy to explain the variations between different applications.

The Commission now strives to make its system more coherent and comprehensible, while recognising the need for stability in international and national regulations, many of which have relatively recently implemented the 1990 recommendations. However, new scientific data have been produced since 1990 and there are developments in societal expectations, both of which will inevitably lead to some changes in the formulation of the recommendations.

---

[1] Presented by Professor Roger Clarke, ICRP Chairman. E-mail roger.clarke@nrpb.org

0952-4746/03/020129+14$30.00 © 2003 IOP Publishing Ltd   Printed in the UK

The previous 1977 recommendations were made in Publication 26 (ICRP 1977), which established the three principles of the system of dose limitation as *justification, optimisation* and *limitation*. Assessments of the effectiveness of protection can be related to the source that gives rise to the individual doses (source-related) or related to the individual dose received by a person from all the sources under control (individual-related). Optimisation of protection is a source-related procedure, while the individual-related dose limits provide the required degree of protection from all the controlled sources.

Optimisation of protection was to be applied to a source in order to determine that doses are 'as low as reasonably achievable, social and economical considerations being taken into account', and decision-aiding techniques were proposed. In particular, the Commission recommended cost–benefit analysis as a procedure to address the question, 'How much does it cost and how many lives are saved?' The Commission recommended that the quantity 'collective dose' should be used in applying those optimisation techniques to take account of the radiation detriment attributable to the source in question. This quantity was unable to take account of the distribution of the individual doses attributable to the source. Attempts were made to address this problem in Publications 37 and 55 (ICRP 1983, 1989), by suggesting a costing of unit collective dose that increased with individual dose received; the procedure was essentially never adopted internationally.

The issue was partially resolved in the 1990 recommendations: while it was still stated, as in 1977, that in relation to any particular source within a practice, the doses should be as low as reasonably achievable, social and economic factors being taken into account, it then continued:

This procedure should be constrained by restrictions on the doses to individuals (*dose constraints*), or the risks to individuals in the case of potential exposures (risk constraints), so as to limit the inequity likely to result from the inherent economic and social judgements (ICRP 1991, paragraph 112).

The concept of the constraint has not been clearly explained by the Main Commission in its subsequent publications. It has not been understood and, although it has been the subject of debate by international bodies, it has not been sufficiently utilised nor has it been implemented widely. The Commission now aims to clarify the meaning and use of the constraint.

The dose constraint was introduced because of the need to restrict the inequity of any collective process for offsetting costs and benefits when this balancing is not the same for all the individuals affected by a source. Before 1990, the dose limit provided this restriction, but in Publication 60 (ICRP 1991) the definition of a dose limit was changed to mean the boundary above which the consequential risk would be deemed unacceptable. This was then considered to be inadequate as the restriction on optimisation of protection and lower value constraints were required to achieve this.

This introduction of the constraint recognised the importance of restricting the optimisation process with a requirement to provide a basic minimum standard of protection for the individual.

The principles for intervention set out in Publication 60 (ICRP 1991) are expressed in terms of a level of dose or exposure where intervention is almost certainty warranted (i.e. justified), which is followed by a requirement to maximise the benefit of the intervention (i.e. the protection level should be optimised). This is effectively an optimisation process and therefore it may be seen in exactly the same terms as for practices, i.e. there is a restriction on the maximum individual dose and then the application of the optimisation process that is itself expected to lead to lower doses to individuals.

It can be seen then that all of the Commission's recommendations since 1990, both for practices and for interventions, have been made in terms of an initial restriction on the

Table 1. ICRP recommendations made since Publication 60 (ICRP 1991).

Publication 62 (ICRP 1993a)	Radiological protection in biomedical research
Publication 63 (ICRP 1993b)	Principles for intervention for protection of the public in a radiological emergency
Publication 64 (ICRP 1993c)	Protection from potential exposure: a conceptual framework
Publication 65 (ICRP 1994)	Protection against radon-222 at home and at work
Publication 75 (ICRP 1997a)	General principles for radiation protection of workers
Publication 76 (ICRP 1997b)	Protection from potential exposures: application to selected radiation sources
Publication 77 (ICRP 1998a)	Radiological protection policy for the disposal of radioactive waste
Publication 81 (ICRP 1998b)	Radiation protection recommendations as applied to the disposal of long-lived solid radioactive waste
Publication 82 (ICRP 1999)	Protection of the public in situations of prolonged radiation exposure

maximum individual dose in the situation being considered, followed by a requirement to optimise protection. This underlines the shift in emphasis to include the recognition of the need for individual protection from a source.

The new recommendations should be seen, therefore, as extending the recommendations in Publication 60 (ICRP 1991), and those published subsequently, to give a single unified set that can be simply and coherently expressed. The opportunity is also being taken to include a coherent philosophy for natural radiation exposures and to introduce a clear policy for radiological protection of the environment.

2. The present situation

Since the 1990 recommendations there have been nine publications, listed in table 1, that have provided additional recommendations for what are effectively to be regarded as 'constraints' in the control of exposures from radiation sources. When ICRP 60 is included, there exist nearly 30 different numerical values for 'constraints', which are set out in table 2, in the ten reports that define current ICRP recommendations. Further, the numerical values are justified in some six different ways, which include:

(a) individual annual fatal risk,
(b) upper end of an existing range of naturally occurring values,
(c) multiples or fractions of natural background,
(d) formal cost–benefit analysis,
(e) qualitative, non-quantitative, reasons, and
(f) avoidance of deterministic effects.

The rationale for the constraints in table 2 is indicated using these letters (a)–(f).

The Commission had previously suggested the term 'protective action level' (PAL) be used in the specification of the restriction of individual doses from single sources. The term appeared to cause concern and was not well understood. The Commission has considered the issue and now feels that the already established term 'constraint' correctly reflects the concept it wishes to promote.

The question to be addressed is whether, for the future, fewer constraints may be recommended that are sufficient to encompass the needs of radiological protection, and whether they can be established on a more uniform and consistent basis.

Table 2. Compilation of the existing ICRP 'constraints' to optimisation: (a) Individual annual fatal risk, (b) upper end of an existing range of naturally occurring values, (c) multiples or fractions of natural background, (d) formal cost–benefit analysis, (e) qualitative, non-quantitative, reasons, and (f) avoidance of deterministic effects.

		Situation to which it applies	
Effective dose[a]	Basis		Publication
		Normal operation of a practice	
~0.01 mSv a^{-1}	a, c	Exemption level, protection optimised	64, 76
0.1 mSv a^{-1}	e	Constraint for long-lived nuclides	82
0.3 mSv a^{-1}	e	Maximum public constraint	77
20 mSv a^{-1}	a	Maximum worker constraint	60, 68
10 mSv a^{-1}	b	Worker constraint for Rn-222	65
(1500 Bq m^{-3})		—optimised level between 500 and 1500 Bq m^{-3}	
2 mSv	e	Surface of the abdomen of pregnant worker	60
1 mSv	a, c	Foetal dose over remaining term of pregnancy	75
1 mSv a^{-1}	a, c	Dose limit for the public	60
		Prolonged exposure	
~10 mSv a^{-1}	c	Below this intervention is optional, but not likely to be justifiable	82
~100 mSv a^{-1}	c, f	Intervention almost always warranted	82
10 mSv a^{-1}	b	Constraint for Rn-222 at home	65
(600 Bq m^{-3})		—optimised level 200–600 Bq m^{-3}	
~1 mSv a^{-1}	c	Intervention exemption level, protection optimised	82
10^{-5}a^{-1}	a	Risk constraint	81
		Biomedical research	
0.1 mSv	a	Minor level of societal benefit	62
1.0 mSv	a	Intermediate level of societal benefit	62
10.0 mSv	a	Moderate level of societal benefit	62
>10.0 mSv	a	Substantial level of societal benefit	62
		Single events and accidents	
Effective dose[a] averted			
50 mSv	e, c	Sheltering warranted—optimised 5–50 mSv	63
500 mSv	e, c	Evacuation warranted—optimised 50–500 mSv	63
5000 mSv skin			
5000 Gy thyroid	e, c	Issue stable iodine—optimised 50–500 mSv	63
1000 mSv	d, a	Arrange relocation (~10 s mSv/month)	63, 82
1000 mSv	f	Constraint for planned emergency work	63
5000 mSv skin			
10 mSv	c, d	Optimised value for foodstuffs 10–100 Bq g^{-1} (α),	63
100 Bq g^{-1} (α),		1000–10 000 Bq g^{-1} (β/γ)	
10 000 Bq g^{-1} (β/γ)			

[a] Unless otherwise stated.

3. Major changes from the 1990 recommendations

The primary aim of the Commission continues to be contributing to the establishment and application of an appropriate standard of protection for human beings and now explicitly for other species. This is to be achieved without unduly limiting those desirable human actions and lifestyles that give rise to, or increase, radiation exposures.

This aim cannot be achieved solely on the basis of scientific data, such as those concerning health risks, but must include consideration of the social sciences. Ethical and economic aspects have also to be considered. All those concerned with radiological protection have to make value judgements about the relative importance of different kinds of risk and about the balancing of risks and benefits. In this, they are no different from those working in other fields concerned with the control of hazards. The restated recommendations will need to recognise this explicitly.

Where exposures can be avoided, or controlled by human action, there is a requirement to provide an appropriate minimum, or basic, standard of protection both for the exposed individuals and for society as a whole. There is a further duty, even from small radiation exposures with small risk, to take steps to provide higher levels of protection when these steps are effective and reasonably practicable. Thus, while the primary emphasis is now on protection of individuals from single sources, it is then followed by the requirement to optimise protection to achieve the best level of protection available under the prevailing circumstances.

In order to achieve this, it is proposed that the existing concept of a constraint be extended to embrace a range of situations to give the levels that bound the optimisation process for a single source. The optimisation of protection from the source may involve either, or both, the design of the source or modification of the pathways leading from the source to the doses in individuals. They would replace a range of terms that include intervention levels and action levels since there would be no need to distinguish intervention situations separately, i.e. constraints, clearance levels and exemption levels, as well as the dose limits for workers and the public.

There will be a revision to the radiation and tissue weighting factors in the definition of effective dose. A coherent philosophy for natural radiation exposures will be included and a clear policy for radiological protection of the environment will be introduced.

4. The 2005 system of protection

The Commission now recognises that there is a distribution of responsibilities for introducing a new source leading to exposures, which lies primarily with society at large, but is enforced by the appropriate authorities. This requires application of the principle of *justification*, so as to ensure an overall net benefit from the source. Decisions are made for reasons that are based on economic, strategic, medical and defence, as well as scientific, considerations. Radiological protection input, while present, is not always the determining feature of the decision and in some cases plays only a minor role. The Commission now intends to apply the system of protection to practices only when they have been declared justified, and to natural sources that are controllable.

The justification of patient diagnostic exposures is included, but has to be treated separately in the recommendations because it involves two stages of decision-making. Firstly, the generic procedure must be justified for use in medicine and, secondly, the referring physician must justify the exposure of the individual patient in terms of the benefit to that patient. It is then followed by a requirement to optimise patient protection, and the Commission has advocated the specification of diagnostic reference levels as indicators of good practice (see section 10).

The system of protection being developed by the Commission is based upon the following principles, which are to be seen as a natural evolution of, and as a further clarification of, the principles set out in Publication 60. Once the source is justified by those appropriate authorities, the radiological principles may be expressed as:

For each source, basic standards of protection are applied for the most exposed individuals, which also protect society—*constraints.*

If the individual is sufficiently protected from a source, then society is also protected from that source.

> However, there is a further duty to reduce doses, so as to achieve a higher level of protection when feasible and practicable. This leads to authorised levels—*optimisation*.

This system of protecting individuals and groups is intended to provide a more coherent basis for protection than the previous one. A necessary basic standard of protection from each relevant source is achieved for individuals by setting constraints that are values of quantities, usually dose, but may be activity concentrations. Constraints are usually annual values, but may be a single value depending on the circumstances.

These constraints or basic levels of protection can be recommended by ICRP and accepted internationally. The responsibility for optimisation then rests with the operators and the appropriate national authority. The operator is responsible for day-to-day optimisation and also for providing input to the optimisation that will establish authorised levels for the operation of licensed practices. These levels will, of necessity, be site and facility dependent and beyond the scope of ICRP.

5. Factors in the choice of new constraints

The present system, which is unduly complex and has used at least six different methods to determine the numerical values, has set maximum constraints that are, in general, about ten times the global average natural background. It is at around this level that doses are usually deemed to require some action, whether they are for practices or intervention, workers or the public.

The Commission now considers the starting point for selecting the levels at which any revised constraints are set to be the concern that can reasonably be felt about the annual dose from natural sources. The existence of natural background radiation provides no justification for additional exposures, but it can be a benchmark for judgement about their relative importance. The worldwide average annual effective dose from all natural sources, including radon, as reported by UNSCEAR is 2.4 mSv (UNSCEAR 2000).

A general scheme for the degree of concern and the level of exposure, as a fraction or multiple of the average annual natural background, is shown in table 3. The fact that the effective dose from natural background varies by at least a factor of ten around the world, and even more if the highest radon doses are included, supports the view that concern should begin to be raised at the higher end of the natural range.

At even higher levels of individual effective dose, i.e. more than 100 mSv in a year, the risk from a source cannot be justified, except in extraordinary circumstances such as life-saving measures in accidents, or in manned space flights. Individual doses of the order of 500 mSv, if acute, can cause early deterministic effects, or, if either acute or delivered over decades, can cause significant probability of increased cancer risk. This then becomes an individual-related restriction on dose and the appropriate authorities must ensure that the individual is not likely to receive significant additional dose from other controllable sources.

At the other extreme, additional effective doses far below the natural background effective annual dose should not be of concern to the individual. Provided that the additional sources come from practices that have not been judged to be frivolous, these doses should also be of no concern to society. If the effective dose to the most exposed is, or will be, less than about 0.01 mSv in a year, then the consequent risk is negligible and protection may be assumed to be optimised, thus requiring no further regulatory concern.

Table 3. Levels of concern and individual effective dose received in a year. Global average annual natural background effective dose from all sources is 2.4 mSv (UNSCEAR 2000).

High	More than 100 mSv
Raised	More than a few tens millisievert
Low	1–10 mSv
Very low	Less than 1 mSv
None	Less than 0.01 mSv

In the intermediate region, doses between a fraction of a millisievert and a few tens of millisievert, whether they are received either singly or repeatedly, are legitimate matters for concern, calling for action by regulatory bodies.

The challenge is whether fewer numbers could replace the 20–30 numerical values for constraints currently recommended in table 2. Further, could they also be more coherently explained in terms of multiples and fractions of natural background.

6. Optimisation of protection

The Commission wishes to retain the phrase 'optimisation of protection' and applies it both to single individuals and to groups. However, it is applied only after meeting the restrictions on individual dose defined by the relevant constraint. It is now used as a short description of the process of obtaining the best level of protection from a single source, taking account of all the prevailing circumstances.

The Commission stated in Publication 77 (ICRP 1998a) that the previous procedure had become too closely linked to formal cost–benefit analysis. The product of the mean dose and the number of individuals in a group, the collective dose, is a legitimate arithmetic quantity, but is of limited utility since it aggregates information excessively. For making decisions, the necessary information should be presented in the form of a matrix, specifying the numbers of individuals exposed to a given level of dose and when it is received. This matrix should be seen as a 'decision-aiding' technique that allows different weightings of their importance to be assigned to individual elements of the matrix. The Commission intends that this will avoid the misinterpretation of collective dose that has led to seriously misleading predictions of deaths.

The concept of collective dose was also previously used as a means of restricting the uncontrolled build-up of exposure to long-lived radionuclides in the environment at a time when it was envisaged that there would be a global expansion of nuclear power reactors and associated reprocessing plants. Restriction of the collective dose per unit of practice can set a maximum future global *per caput* annual effective dose from all sources under control. If, at some point in the future, a major expansion of nuclear power were to occur, then some reintroduction of a procedure may have to be considered to restrict a global build-up of *per caput* dose.

The process of optimisation may now be expressed in a more qualitative manner. On a day-to-day basis the operator is responsible for ensuring the optimum level of protection and this can be achieved by all those involved, workers and professionals, always challenging themselves as to whether protection can be improved. Optimisation is a frame of mind, always questioning whether the best has been done in the prevailing circumstances. For the more formal authorisations, which are decided by the regulator in conjunction with the operator, they may in future best be carried out by involving all the bodies most directly concerned, including representatives of those exposed, in determining, or in negotiating, the best level of protection in the circumstances. It is to be decided how the Commission's recommendations will deal with this degree of societal process. However, the result of this process will lead to the authorised levels applied by the regulator to the source under review.

7. Exclusion of sources and exposures

The Commission intends its system of protection to apply to the deliberate introduction of a new controllable source or the continued operation of a controllable source that has deliberately been introduced, i.e. a practice, and to controllable natural sources. Its recommendations can then be applied to reduce doses, when either the source or the pathways from the source to the exposed individuals can be controlled by some reasonable means. Sources that do not fall within this definition of controllable are excluded from regulatory control. There are sources for which the resulting levels of annual effective dose are very low, or for which the difficulty of applying controls is so great and expensive that protection is already optimised and the sources are therefore excluded.

In its restated policy the Commission defines what sources and exposures are to be excluded from the system of protection and will not use the term 'exemption'. Exemption or clearance is seen as a regulatory decision that is applied to non-excluded sources by the appropriate regulatory body. That body has the responsibility for deciding when radioactive material is to be released from its control, which is in effect an 'authorised release' no different from that specified for effluent discharges after application of the optimisation process.

Apart from these exclusions, the Commission has aimed to make its recommendations applicable as widely and as consistently as is possible, irrespective of the origin of the sources. The Commission's recommendations thus will now cover exposures to both natural and artificial sources, so far as they are controllable.

8. Natural sources

The Commission intends to include explicit recommendations for protection from natural radiation sources. It is clear that it is the controllability of the exposure that determines whether the exposures are excluded from, or included in, the system of protection. In particular, the control of radon-222 is a special case because of its ubiquitous nature.

The ICRP recommendations for radon-222 in Publication 65 (ICRP 1994) have been widely accepted and the Commission proposes they should continue. These suggested a maximum level of dose (the constraint) that was translated into an activity concentration, then followed by an activity concentration range within which an optimised 'action level' would be found (table 2). As now, the recommendation would be that for exposures above this level, the system of protection is applied. Exposures below the designated level are then excluded from the system of protection. The Commission now refers to this designated level as the *exclusion level*.

The Commission is now considering an approach analogous to that for radon-222 for protection from the other controllable natural sources. The principal sources of both internal and external exposure in environmental materials are potassium-40 and the decay series of uranium-238 and thorium-232. The Commission is considering recommending a maximum constraint for these, on the grounds that it is impractical to control all natural sources. The constraint, as with radon, would not be expressed in dosimetric quantities but rather as an activity concentration, since that is more appropriate and with a value at the upper end of the existing natural range. The appropriate authority would apply generic optimisation, or broad experience on practicability, to find an *exclusion level* that is lower than the constraint.

The only protective actions are relocation of populations and, if the sources are mainly in building materials, extensive rebuilding. These actions are disruptive and require considerable resources. Thus the exclusion level, while lower than the constraint, but probably not by more than a factor of a few, will be somewhere in the naturally existing range, corresponding probably to dose of about a fraction of a millisievert annually.

Cosmic rays at ground level and the resultant exposures are not controllable. They are therefore excluded from the scope of the recommendations. Limiting the time spent by passengers and crew at high altitudes is the only action that could control exposure to cosmic rays in aircraft. The average annual effective doses to some aircrew are about 3 mSv, while the exposure of some specialist aircrew and a few professional couriers may be twice as high. The Commission has recommended in Publication 60 (ICRP 1991) that the exposures of aircrew in the operation of jet aircraft should be treated as occupational exposure in its system of protection. The Commission considers that there is no justification for controlling doses to members of the public from flying and these exposures should be excluded.

9. Dosimetric quantities

There have been some persistent difficulties with, and misunderstandings of, the definitions of the Commission's dosimetric quantities. The Commission will remove these by clarifying its definitions and specifying their application.

The Commission uses the *weighted averaged absorbed* dose in an organ or tissue. It no longer uses the term 'equivalent dose' in order to avoid confusion with 'dose equivalent' in translation to other languages. The implicit averaging is valid only if the range of doses is such that the proportional dose–effect relationship applies. There is no proposal to move away from the use of effective dose as currently defined,

$$E = \sum_T w_T \sum_R w_R D_{T,R}.$$

There is, however, a need to reconsider the basis used to derive the numerical values for both the tissue and radiation weighting factors. There is evidence from a recent report by a Task Group of ICRP Committee 1 that the w_R values for protons and neutrons may need revision (ICRP 2003a). The UNSCEAR report (2001) gives reduced estimates of the risk of hereditary defects and another Task Group of Committee 1 is developing proposals to simplify the way in which cancer risks are used to establish w_T values.

It should be emphasised that effective dose is intended for use as a protection quantity and therefore should not be used for epidemiological evaluations, nor should it be used for any specific investigation of human exposure. Rather, absorbed dose should be used with the most appropriate biokinetic, biological effectiveness and risk factor data.

Those health effects that exhibit a dose threshold result from the loss of function of a significant number of cells in a tissue. The dosimetric situation causing this loss of function is complex. If the dose is approximately uniform over the tissue, the mean absorbed dose is a good starting point. If the dose is far from uniform, the localised damage may not reduce the performance of the tissue, but the localised damage may be severe. All these situations depend heavily on the distributions of delivered dose in position and time. This complexity cannot be reflected in the dosimetric quantities. The only approach is to make qualitative judgements based on the distribution of absorbed dose in location and time. In many cases, there is no need to introduce any weighting based on RBE, because its value will rarely exceed two.

10. Radiation exposure of patients

The application to the medical uses of radiation for patients requires separate guidance. Limitation of the dose to the individual patient is not recommended as it may, by reducing the effectiveness of the patient's diagnosis or treatment, do more harm than good. The emphasis is then on the justification of the medical procedures.

There are three levels of justification of a procedure in medicine. At the first and most general level, the use of radiation in medicine is accepted as doing more good than harm. Its justification is now taken for granted. At the second level, a specified procedure with a specified objective is defined and justified, e.g. chest radiographs for patients showing relevant symptoms. The aim of this generic justification is to judge whether the radiological procedure will usually improve the diagnosis or treatment or will provide necessary information about the exposed individuals. At the third level, the application of the procedure to an individual patient should be justified, i.e. the particular application should be judged to do more good than harm to the individual patient.

The subsequent formal optimisation concentrates on the requirement to keep the doses to patients as low as is consistent with the medical objectives. In diagnosis this means reducing unnecessary exposures, while in therapy it requires delivery of the required dose to the volume to be treated, avoiding exposure of healthy tissues.

The generic justification of a defined medical procedure

The generic justification of the procedure is a matter for national professional bodies, sometimes in conjunction with national regulatory authorities. The total benefits from a medical procedure include not only the direct health benefits to the patient, but also the benefits to the patient's family and to society. Although the main exposures in medicine are to patients, the exposures to staff and to members of the public who are not connected with the procedures should be considered. The possibility of accidental or unintended exposures (potential exposure) should also be considered. The decisions should be reviewed from time to time, as more information becomes available about the risks and effectiveness of the existing procedure and about new procedures.

The justification of a procedure for an individual patient

For complex diagnostic procedures and for therapy, generic justification may not be sufficient. Individual justification by the radiological practitioner and the referring physician is then important and should take account of all the available information. This includes the details of the proposed procedure and of alternative procedures, the characteristics of the individual patient, the expected dose to the patient, and the availability of information on previous or expected examinations or treatment.

Diagnostic reference levels

These are used in medical diagnosis to indicate that, in routine conditions, the dose to the patient from a specified procedure should not normally exceed the reference level for that procedure as indicated by a measurable quantity such as entry dose in an x-ray examination. These have already been used as guidance levels for medical diagnostic procedures in the IAEA Basic Safety Standards and in the Euratom Directive on health protection against ionising radiation in medical exposure.

11. Radiological protection of the living environment

The current ICRP position regarding protection of the environment is set out in Publication 60 (ICRP 1991): *'The Commission believes that the standards of environmental control needed to protect man to the degree currently thought desirable will ensure that other species are not put at risk.'* Up until now, the ICRP has not published any recommendations as to how protection of the environment should be carried out. The Commission has recently adopted a report dealing with environmental protection (ICRP 2003b). This report addresses the role that ICRP could play in this important and developing area, building on the approach that has been developed for human protection and on the specific area of expertise to the Commission, namely radiological protection.

The Commission has decided that a systematic approach for radiological assessment of non-human species is needed in order to provide the scientific basis to support the management of radiation effects in the environment. This decision to develop a framework for the assessment of radiation effects in non-human species has not been driven by any particular concern over environmental radiation hazards. It has rather been developed to fill a conceptual gap in radiological protection and to clarify how the proposed framework can contribute to the attainment of society's goals of environmental protection by developing a protection policy based on scientific and ethical–philosophical principles.

The proposed system does not intend to set regulatory standards. The Commission rather recommends a framework that can be a practical tool to provide high-level advice and guidance and help regulators and operators demonstrate compliance with existing legislation. The system does not preclude derivation of standards, on the contrary it provides a basis for such derivation.

At present, there are no internationally agreed criteria or policies that explicitly address protection of the environment from ionising radiation, although many international agreements and statutes call for protection against pollution generally, including radiation. The current system of protection has indirectly provided protection of the human habitat. The lack of a technical basis for assessment, criteria or standards that have been endorsed at an international level makes it difficult to determine or demonstrate whether or not the environment is adequately protected from potential impacts of radiation under different circumstances. The Commission's decision to develop an explicit assessment framework will support and provide transparency to the decision-making process.

A framework for radiological protection of the environment must be practical and simple. The ICRP framework will be designed so that it is harmonised with its proposed approach for the protection of human beings. To achieve this, an agreed set of quantities and units, a set of reference dose models, reference dose-per-unit-intake data and effects-analysis will be developed. A limited number of reference fauna and flora will be developed by ICRP to aid assessments, and others can then develop more area- and situation-specific approaches to assess and manage risks to non-human species. ICRP has a unique position in relation to human radiological protection, from which it has played a major role in influencing legal frameworks and objectives at international and national levels. In contrast, the subject of protection of other species is more complex and multifaceted, with many international and national environmental legislative frameworks and objectives already in place.

The Commission proposes that the objectives of a common approach to the radiological protection of non-humans organisms are to safeguard the environment by preventing or reducing the frequency of effects likely to cause early mortality or reduced reproductive success in individual fauna and flora to a level where they would have a negligible impact on conservation of species, maintenance of biodiversity, or the health and status of natural habitats or communities.

A considerable challenge for ICRP will be that of integrating any approach to protection of the environment with that of the protection of human beings, bearing in mind that the latter is also the subject of a current, in-depth, review. ICRP can, and is prepared, to play the key role with respect to ionising radiation in the environment, both in advising on a common international approach, and in providing the basic interpretation of existing scientific information. This will include identifying where further research is necessary—in order for such a common approach to be delivered.

12. Some outstanding issues and proposed timescales

The Main Commission is preparing a number of supporting documents on which the main recommendations will draw. These include summaries of the health effects of radiation at low doses and the review of RBE values, which together will lead to a document on the decision for revised radiation and tissue weighting factors. Other major issues which are under development and need further discussion are:

- Exploration into the possibility of specifying fewer numerical constraints than currently exist and whether they can be more coherently explained.
- Clarification of the exclusion concept and further elaboration of the observation that all releases from regulatory control are 'authorised releases'.
- A review of the 'critical group' concept as used to represent the hypothetical individual. ICRP has not addressed this since well before the 1990 recommendations.
- Development of methods by which the optimisation of protection can realistically be achieved.

The intention is to have draft recommendations prepared for discussion with the four Committees late in 2003 so that a well-developed draft is available for the IRPA 11 Congress in May 2004. It is planned to produce the final version in 2005. Table 4 shows a brief compilation of some of the major topics where there will be changes from present recommendations to the new proposals.

Résumé

Au cours de ces dernières années, la CIPR a encouragé les discussions sur la meilleure façon d'exprimer la philosophie de radioprotection dans ses prochaines Recommandations qu'elle prévoit de publier en 2005. Les recommandations actuelles ont été établies dans le cadre de la Publication 60 en 1990, puis ont été complétées par d'autres publications au cours des douze dernières années. Il est à présent évident que la Commission doit rassembler la totalité des valeurs numériques qu'elle a recommandées dans ces différents rapports. C'est ce qui est fait dans le présent document, et sur ces bases est proposée une façon de progresser vers un système de radioprotection simplifié et plus cohérent pour le début du XXIème siècle. Plutôt qu'une révision radicale, ce qui est envisagé c'est une présentation cohérente de la politique actuelle ainsi qu'une simplification de son application.

Zusammenfassung

Während der letzten Jahre hat die ICRP zu Diskussionen ermuntert über den besten Weg, um die Grundgedanken (Philosophie) des Strahlenschutzes in den nächsten Empfehlungen zum Ausdruck zu bringen. Diese sollen im Jahre 2005 veröffentlicht werden. Die gegenwärtigen Empfehlungen wurden 1990 durch die ICRP-Veröffentlichung Nr. 60 initiiert. Die Publikation

Table 4. Brief summary of essential changes expected in the new recommendations.

Topic	Present recommendations	New recommendations
Linearity	Linear non-threshold, i.e. proportionality	Clarify concept and applicable range, i.e. above a few millisievert per year
Effective dose	Yes	Yes
Radiation weighting factor	Publication 60	Revised values for protons and neutrons
Tissue weighting factor	Publication 60	New values based on revised risk factors and a simplified basis
Nominal risk coefficient	Publication 60	Total cancer fatality similar, but individual organs changed hereditary use (UNSCEAR 2001)
Limits	Worker and public in Publication 60	Incorporated into revised constraints
Constraints	See table 2	Number and complexity to be reduced
Collective dose	Publication 60	Disaggregated and replaced by weighted matrix
Justification	Publication 60	Retained, extended for patient exposure
Optimisation	Cost–benefit analysis	Stakeholder involvement
Exemption	Publication 60	Replace by exclusion
Definition of 'individual'	Publication 29	New consideration
Practice	Publication 60	Retain
Intervention	Publication 60	Incorporate into constraints
Environment (non-human)	Assumed protected in Publication 60	Explicitly addressed
Natural radiation sources	Radon-222 only	Comprehensive treatment

wurde ergänzt durch zusätzliche Berichte während der letzten zwölf Jahre. Es ist jetzt klar geworden, dass für die Kommission die Notwendigkeit besteht, die Gesamtheit der zahlreichen numerischen Werte zusammenzufassen, die sie in mehr als zehn Berichten empfohlen hat. Dieses ist in der vorliegenden Arbeit geschehen. Hiervon ausgehend wird ein Weg aufgezeigt, wie eine vereinfachte und kohärentere Darstellung der Schutz-Philosophie zu Beginn des 21. Jahrhunderts erreicht werden kann. Eine radikale Revision wird nicht ins Auge gefasst sondern vielmehr eine verständliche Darlegung der gegenwärtigen Strukturen und eine Vereinfachung ihrer Anwendung.

References

ICRP 1977 Recommendations of the ICRP on radiological protection *ICRP Publication* 26 (*Ann. ICRP* **1** (3))

ICRP 1983 Cost-benefit analysis in the optimization of radiation protection *ICRP Publication* 37 (*Ann. ICRP* **10** (2/3))

ICRP 1989 Optimization and decision-making in radiological protection *ICRP Publication* 55 (*Ann. ICRP* **20** (1))

ICRP 1991 1990 Recommendations of the ICRP *ICRP Publication* 60 (*Ann. ICRP* **21** (1–3))

ICRP 1993a Radiological protection in biomedical research *ICRP Publication* 62 (*Ann. ICRP* **22** (3))

ICRP 1993b Principles for intervention for protection of the public in a radiological emergency *ICRP Publication* 63 (*Ann. ICRP* **22** (4))

ICRP 1993c Protection from potential exposure: a conceptual framework *ICRP Publication* 64 (*Ann. ICRP* **23** (1))

ICRP 1994 Protection against radon-222 at home and at work *ICRP Publication* 65 (*Ann. ICRP* **23** (2))

ICRP 1997a General principles for the radiation protection of workers *ICRP Publication* 75 (*Ann. ICRP* **27** (1))

ICRP 1997b Protection from potential exposures: an application to selected radiation sources *ICRP Publication* 76 (*Ann. ICRP* **27** (2))

ICRP 1998a Radiological protection policy for the disposal of radioactive waste *ICRP Publication* 77 (*Ann. ICRP* **27** suppl)

ICRP 1998b Radiation protection recommendations as applied to the disposal of long-lived solid radioactive waste *ICRP Publication* 81 (*Ann. ICRP* **28** (4))

ICRP 1999 Protection of the public in situations of prolonged radiation exposure *ICRP Publication* 82 (*Ann. ICRP* **29** (1/2))

ICRP 2003a Relative biological effectiveness (RBE), radiation weighting factor (w_R), and quality factor (Q) *ICRP Publication* 92 (*Ann. ICRP* **33** (4))

ICRP 2003b A framework for assessing the impact of ionizing radiation on non-human species *ICRP Publication* 91 (*Ann. ICRP* **33** (3))

UNSCEAR 2000 Sources and effects of ionizing radiation *2000 Report to the General Assembly* (New York: United Nations) p E.00.IX.4

UNSCEAR 2001 Hereditary effects of radiation *2001 Report to the General Assembly* (New York: United Nations) p E.01.IX.2

2005 RECOMMENDATIONS OF THE INTERNATIONAL COMMISSION ON RADIOLOGICAL PROTECTION

SUMMARY OF THE RECOMMENDATIONS

(S1) This Summary indicates the Commission's aims and the way in which the recommendations may be applied. The necessary concepts are defined and explained in the main text following this Summary.

The Aim of the Recommendations

(S2) The fundamental aim of the Commission was set out as follows in the 1990 Recommendations.

> 'The primary aim of radiological protection is to provide an appropriate standard of protection for man without unduly limiting the beneficial actions giving rise to radiation exposure. This aim cannot be achieved on the basis of scientific concepts alone. All those concerned with radiological protection have to make value judgements about the relative importance of different kinds of risk and about the balancing of risks and benefits. In this, they are no different from those working in other fields concerned with the control of hazards.'

This statement still represents the Commission's position.

(S3) The Commission has concluded that its recommendations should be based on a simple, but widely applicable, general system of protection that will clarify its objectives and will provide a basis for the more formal systems needed by operating managements and regulators. It also recognises the need for stability in regulatory systems at a time when there is no major problem identified with the practical use of the present system of protection in normal situations. The use of the optimisation principle, together with the use of constraints and the current dose limits, has led to a general overall reduction in both occupational and public doses over the past decade. The Commission now strengthens its recommendations by quantifying constraints for all controllable sources in all situations.

The Principles of Protection

(S4) The system of protection now recommended by the Commission is to be seen as a natural evolution of, and as a further clarification of, the 1990 Recommendations. The 2005 Recommendations establish quantified restrictions on individual dose from specified sources in all situations within their scope. These restrictions should be applied to the exposure of actual or representative individuals. They provide a level of protection for individuals that should be considered as obligatory, and not maintaining these levels of protection should be regarded as a failure. The quantified restrictions are complemented by the requirement to optimise the level of protection achieved.

(S5) The most fundamental level of protection is the source-related restriction on individual dose called a *dose constraint*. It is used to provide a level of protection for the most exposed *individuals* within a class of exposure, in all situations within the scope of the recommendations, *from a single source*. Except for the exposure of patients, these constraints should be regarded as the basic levels of protection to be attained in all situations that are addressed by the Commission: normal situations, accidents and emergencies, and the case of controllable existing exposure. These constraints represent the level of dose where action to avert exposures and reduce doses is virtually certain to be justified.

(S6) In all situations the constraints are complemented by the requirement to optimise the level of protection achieved. This is because there is presumed to be some probability of health effects even at small increments of exposure to radiation above the natural background. The Commission therefore recommends that further, more stringent, measures should be considered for each individual source. This requirement for the optimisation of protection includes, but is more comprehensive than, the need to ensure that all exposures are as low as reasonably achievable, economic and social factors being taken into account, in the relevant situation. This requirement cannot be defined in general quantitative terms; it calls for judgement about each situation causing exposure of individuals and is the concern of the operating managements and the responsible national authorities.

(S7) Table S1 presents the Commission's recommended maximum values of dose constraints. In essence, four values are recommended according to the type of situation to be controlled. They should be considered as giving the upper restriction that is to be applied by the appropriate national authorities to determine the most applicable constraints for the situation under consideration. The Commission expects that the resulting national values of constraints normally will be lower than the maximum value recommended by the Commission, but probably not by as much as a factor of ten.

Table S1. Maximum dose constraints recommended for workers and members of the public from single dominant sources for all types of exposure situations that can be controlled.

Maximum constraint (effective dose, mSv in a year)	Situation to which it applies
100	In emergency situations, for workers, other than for saving life or preventing serious injury or preventing catastrophic circumstances, and for public evacuation and relocation; and for high levels of controllable existing exposures. There is neither individual nor societal benefit from levels of individual exposure above this constraint.
20	For situations where there is direct or indirect benefit for exposed individuals, who receive information and training, and monitoring or assessment. It applies into occupational exposure, for countermeasures such as sheltering, iodine prophylaxis in accidents, and for controllable existing exposures such as radon, and for comforters and carers to patients undergoing therapy with radionuclides.
1	For situations having societal benefit, but without individual direct benefit, and there is no information, no training, and no individual assessment for the exposed individuals in normal situations.
0.01	Minimum value of any constraint

(S8) The level of protection for an individual from all sources within a class of exposure, in normal situations only, is the *dose limit*. The Commission has recommended values of dose limits in its 1990 Recommendations, ICRP *Publication 60*, which have been adopted in international safety standards and in the national legislation of nearly all countries. The Commission continues to recommend the use of its 1990 dose limits, in normal situations only.

Optimisation of Protection

(S9) Optimisation of protection is a process that is an important component of a successful radiological protection programme. In application, it involves evaluating and, where practical to do so, incorporating measures that tend to lower radiation doses to members of the public and to workers. But conceptually it is broader, in that it entails consideration of the avoidance of accidents and other potential exposures. It incorporates a range of qualitative and quantitative approaches.

(S10) An important role of the concept of optimisation of protection is to foster a 'safety culture' and thereby to engender a state of thinking in everyone responsible for control of radiation exposures, such that they are continuously asking themselves the question, 'Have I done all that I reasonably can to reduce these doses?' Clearly, the answer to this question is a matter of judgement and necessitates co-operation between all parties involved and, as a minimum, the operating management and the regulatory agencies.

(S11) The involvement of *stakeholders*, a term which has been used by the Commission in *Publication 82* to mean those parties who have interests in and concern about a situation, is an important input to optimisation. While the extent of stakeholder involvement will vary from one situation to another in the decision-making process, it is a proven means to achieve the incorporation of values into decisions, the improvement of the substantive quality of decisions, the resolution of conflicts among competing interests, the building of trust in institutions as well as the education and information the workers and the public. Furthermore, involving all parties affected by the decision reinforces the safety culture and introduces the necessary flexibility in the management of the radiological risk that is needed to achieve more effective and sustainable decisions.

Exclusion of radiation sources

(S12) There are many sources for which the resulting levels of annual effective dose are very low, or for which the combination of dose and difficulty of applying control are such that the Commission considers that the sources can legitimately be *excluded* completely from the scope of its Recommendations. Since cosmic rays are ubiquitous and all materials are radioactive to a greater or lesser degree, the concept of exclusion is essential for the successful application of the system of protection. The Commission has concluded that the activity concentration values in Table S2 provide a definition of what is to be considered radioactive for practical radiological protection purposes, and therefore the levels at which materials are to be within the scope of its recommendations. It now recommends the figures in Table S2 as the basis of exclusion from the scope of its recommendations.

Table S2. Recommended Exclusion Levels

Nuclides	Exclusion activity concentration
Artificial a -emitters	0.01 Bq g^{-1}
Artificial ß/? emitters	0.1 Bq g^{-1}
Head of chain activity level[†], ^{238}U, ^{232}Th	1.0 Bq g^{-1}
^{40}K	10 Bq g^{-1}

[†] For ^{238}U and ^{232}Th chains, this value also applies to any nuclide in a chain that is not in secular equilibrium excluding ^{222}Rn and daughters in air which in all situations are controlled separately.

The development of effective dose

(S13) The weighting factors in calculating effective dose are intended to take account of many types of radiation, many types of stochastic effects, and many tissues in the body. They are therefore only loosely based on a wide range of experimental data. It is unrealistic to expect them to apply accurately to any particular case. In recent recommendations, the Commission has deliberately selected broadly based values of these weighting factors.

(S14) The weighting factor for radiation quality is applied directly to the absorbed dose in a tissue or organ. This weighted tissue dose has been called both dose equivalent and equivalent dose at various times. There has been substantial confusion between these terms, particularly in translation from English into other languages. The Commission now avoids both of those terms and uses *radiation weighted dose* in a tissue or organ. The unit of radiation weighted dose is the joule per kilogram with the special name sievert (Sv). The Commission is considering a new special name for radiation weighted dose so as to avoid the use of the name 'sievert' for both radiation weighted dose and effective dose.

(S15) When, as is usual, more than one tissue is exposed, it is necessary to use the tissue weighting factor. The application of both the radiation and the tissue weighting factors to the tissue absorbed doses leads to the effective dose. The effective dose, as currently defined, will continue to be used by the Commission for protection purposes,

$$E = \sum_T w_T \sum_R w_R \cdot D_{T,R}$$

where E is the effective dose, w_R and w_T are the radiation and tissue weighting factors, and $D_{T,R}$ is the mean absorbed dose in tissue or organ T due to incident radiation R. The unit of effective dose is the joule per kilogram and called the sievert (Sv). Since the effective dose is derived from mean absorbed doses in tissues and organs of the human body, a dosimetric model must be specified or implied in any statement of the magnitude of the effective dose.

(S16) As in the 1990 Recommendations, radiation weighting factors are determined by the characteristics of the type and energy of the radiation incident on the body or, in the case of sources within the body, emitted by the source. The radiation weighting factors are then applied to the mean tissue dose in any specified part of the human body. The radiation weighting factors in Table S3 are essentially those suggested in *Publication 92* and are now recommended for general use in radiological protection. For neutrons a continuous curve is recommended shown in Figure S1. In order to reduce computational difficulties in evaluating effective dose the function in Figure S1 is given in Equation S1.

$$w_R = \begin{cases} 2.5 + 18.2 \exp[-(\ln E_n)^2/6] & \text{for } E_n < 1 \text{ MeV} \\ 5.0 + 17.0 \exp[-(\ln (2E_n))^2/6] & \text{for } E_n \geq 1 \text{ MeV.} \end{cases} \quad \text{.........................(S1)}$$

where E_n is in MeV. The radiation weighting factor for neutrons is applied to the mean absorbed doses in the relevant tissues and organs. The dose is that from both the neutron induced charged particles and the secondary photons induced in the body.

(S17) The Commission has reviewed the epidemiological data that can be used to assess nominal risk factors for cancer and hereditary diseases. From these it has developed a new estimate of detriment resulting from radiation exposure which has been used to specify its recommended w_T values. The new values that apply for the tissue weighting factors are listed below in Table S4. The weighting factor for Remainder tissues is to be applied to dose averaged over the 14 specified organs and tissues that constitute the Remainder.

Table S3. Radiation weighting factors, w_R

Type and energy range	w_R
Photons	1
Electrons and muons	1
Protons	2
Alpha particles, fission fragments, heavy nuclei	20
Incident neutrons	**See Figure S1 and Equation S1**

Figure S1. Radiation weighting factor, w_R, for incident neutrons versus neutron energy. (A) Step function and (B) continuous function given in *Publication 60*, (C) function proposed in this report.

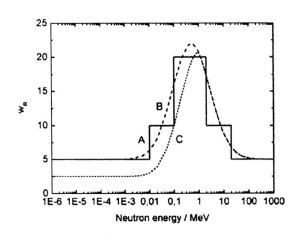

Neutron energy / MeV

Table S4. Tissue weighting factors

Tissue	w_T	? w_T
Bone marrow, Breast, Colon, Lung, Stomach	**0.12**	0.60
Bladder, Oesophagus, Gonads, Liver, Thyroid	**0.05**	0.25
Bone surface, Brain, Kidneys, Salivary glands, Skin	**0.01**	0.05
Remainder Tissues*	**0.10**	0.10

*Remainder Tissues (14 in total)

Adipose tissue, Adrenals, Connective tissue, Extrathoracic airways, Gall bladder, Heart wall, Lymphatic nodes, Muscle, Pancreas, Prostate, SI Wall, Spleen, Thymus, and Uterus/cervix.

The development of a framework for the protection of non-human species

(S18)   The Commission's new framework for non-human species will be designed so that it is harmonized with its proposed approach for the protection of human beings. To achieve this, an agreed set of nomenclature, plus a set of reference dose models, data sets to rela te exposure to dose, and interpretation of effects will be developed for a limited number of animal and plant types. This will also ensure that the protection of both humans and other organisms are protected on the same scientific basis, in terms of the re lationships between exposures to ionising radiation and dose, and between dose and effects at the molecular, cellular, tissue and organ, and whole organism level.

(S19)   The Commission recognises that a framework for radiological protection of the environment must be practical and, ideally, a set of ambient activity concentration levels would be the simplest tool. There is a need for international standards of discharges into the environment, and the Commission's common approach will provide a basis for the develbpment of such standards. In order to demonstrate, transparently, the derivation of ambient activity concentration levels or standards, the reference-animal-and-plant approach will be helpful.

The Intended Use of the Recommendations

(S20)   The Commission's advice has to be of a general and international nature. However, the Commission hopes that its advice will influence both regulatory agencies and management bodies, including their specialist advisors. It also hopes that its advice will continue to help in the provision of a consistent basis for national and regional regulatory policies and standards. The Commission recognises that these hopes will be fulfilled only if there is general acceptance of its judgements and policies by the managements of practices causing exposures to radiation, by regulatory agencies, and by governments. Its experience since its establishment in 1928 leads the Commission to conclude that this coherent acceptance exists.

(S21)   The Commission aims to provide guidance to a wide range of organisations in a wide range of countries and regions. The Commission believes that these bodies have the responsibility to design their own procedures, which may require development of their own internal documents. The Commission's underlying hope is that it can encourage the widespread development of a radiological safety culture, which lies within the framework of its recommendations, and which then permeates all the operations involving exposure to ionising radiation. The starting point for this should be a programme of relevant education and training.

-- The complete draft is available at www.icrp.org *--*